Intelligent Perception and Information Processing

This book series Intelligent Perception and Information Processing is grounded in artificial intelligence and signal and information processing theory, focusing on key technologies such as communication, vision, and location for signal and information perception, processing, transmission, and storage. It is committed to solving key and difficult problems in wireless perception, machine vision, information and coding, image processing, multi-source information fusion, environment configurable intelligent transmission, cross-media transmission, knowledge graph and other application areas, and promoting the development of intelligent perception and signal processing.

Yao Ding · Zhili Zhang · Haojie Hu · Fang He ·
Shuli Cheng · Yijun Zhang

Graph Neural Network for Feature Extraction and Classification of Hyperspectral Remote Sensing Images

Science Press
Beijing

Springer

Yao Ding
Xi'an Research Institute of High Technology
Xi'an, Shaanxi, China

Haojie Hu
Xi'an Research Institute of High Technology
Xi'an, Shaanxi, China

Shuli Cheng
Xinjiang University
Urumqi, Xinjiang, China

Zhili Zhang
Xi'an Research Institute of High Technology
Xi'an, Shaanxi, China

Fang He
Xi'an Research Institute of High Technology
Xi'an, Shaanxi, China

Yijun Zhang
Experimental Training Base
National University of Defense Technology
Xi'an, Shaanxi, China

ISSN 3059-3808 ISSN 3059-3816 (electronic)
Intelligent Perception and Information Processing
ISBN 978-981-97-8008-2 ISBN 978-981-97-8009-9 (eBook)
https://doi.org/10.1007/978-981-97-8009-9

Jointly published with Science Press
The print edition is not for sale in China mainland. Customers from China mainland please order the print book from: Science Press.

This Springer imprint is published by the registered company Springer Nature Singapore Pte Ltd.
The registered company address is: 152 Beach Road, #21-01/04 Gateway East, Singapore 189721, Singapore

Preface

Remote sensing (RS) is a non-contact Earth observation technology that obtains ground target information through the reflection or radiation of electromagnetic waves on ground objects. Military remote sensing technology can use remote sensing carriers, such as satellites and drones to conduct high-resolution, real-time, multi-angle, and multi-frequency remote sensing observations of enemy military targets, obtain enemy military intelligence and battlefield conditions, providing intelligence support, tactical guidance, and intelligence support for military command departments. This technology plays a crucial role in modern warfare and is of great significance in improving combat efficiency and reducing war costs. Hyperspectral remote sensing images contain high spectral resolution, rich spectral information, and wide band range, which can more accurately identify target features. Extracting and analyzing the spatial–spectral information of hyperspectral images, assigning specific category labels to each pixel in the image based on the different types of covered objects, is an important means of converting hyperspectral image data into usable knowledge and is also the foundation of hyperspectral image analysis and research.

However, in practical applications, there are still many problems that need to be addressed, such as curse of dimensionality, nonlinear data structures, spectral uncertainty, spatial homogeneity, and spatial heterogeneity. How to solve these problems is worth pondering deeply. To fully utilize the rich spectral information contained in hyperspectral images, more and more machine learning and artificial intelligence algorithms are widely used in hyperspectral image classification technology. Due to the unique advantages, graph neural network has received increasing attention, and the combination of hyperspectral image classification and graph learning is one of the current research hotspots. The reasons include: First, graph neural network methods can learn the relationships between nodes and do not require a high number of labeled samples; secondly, graph neural networks can automatically learn and process the feature signals (spectral signals) of nodes. For hyperspectral image with high-dimensional spectral data, graph neural networks have natural advantages that convolutional neural networks do not have; thirdly, the combination of graph neural networks and machine learning methods can effectively distinguish the contours of covered objects and improve classification accuracy.

The main research content of this book is to use graph learning methods for hyperspectral image classification, focusing on model construction, improving graph information dissemination methods, and improving composition quality. Various graph learning-based hyperspectral image classification methods are proposed. The main research content of this book is summarized as follows: Chap. 1 mainly introduces the research background of this book, explains the practical significance, and outlines the current situation and existing problems of hyperspectral remote sensing image classification. To facilitate the understanding of this book, the basic knowledge of graph networks and the evaluation indicators used in the experiments are also introduced. Furthermore, the research content and chapter arrangement of this book are specifically presented. Chapter 2 proposes the graph sample and aggregate attention network to address traditional graph convolutional neural networks having high computational complexity, inability to aggregate new nodes, and susceptibility to over-smoothing as the number of convolutional layers increases. Chapter 3 addresses the difficulties in integrating graph neural networks and convolutional neural networks, as well as the problem of single-pixel classification errors caused by superpixel image preprocessing. Chapter 4 focuses on the computational complexity, over-smoothing, and unable automatically adjust the aggregation path of graph neural network. An adaptive path aggregation graph neural framework is proposed for hyperspectral image classification. Chapter 5 first performs feature learning on large-scale irregular regions through GNN and then extracts local spatial–spectral features at the pixel level. Besides, edge convolution (EdgeConv) is incorporated into GNN to adaptively capture the interrelationship of the representative descriptors and fully exploit the discriminative features on graph. Finally, the graph neural network via edge convolution is proposed for hyperspectral image classification. Chapter 6 focuses on the problem that the constructed graph may not be ideal for the downstream classification task due to the variety of spectral characteristics. First, a fully connected graph is adaptively constructed to make full use of local spatial information and global spectral information. Then, a neural sparsification technique is applied to remove potentially task-irrelevant edges in the case of misleading message propagation. Furthermore, label propagation (LP) serves as regularization to assist the graph network in learning proper edge weights that lead to improved classification performance. Finally, a unifying label propagation and graph sparsification network is proposed for hyperspectral image classification. Chapter 7 proposes the graph-former framework for spatial spectral feature enhancement (S^2GFormer) to address the difficulties in extracting local–non-local spatial spectral features for hyperspectral image classification by comprehensively utilizing the advantages of transformer and GCN. Chapter 8 proposes an end-to-end method for large-scale hyperspectral image clustering, addressing the challenges of training label dependencies with supervised methods.

Chapter 1 of this book was written by Ding Yao, Zhang Zhili, Zhang Yijun, and Yu Chengguo, Chaps. 2–4 were written by Ding Yao, Zhang Zhili, and Yu Chengguo, Chaps. 5–6 were written by Hu Haojie and He Fang, and Chap. 7 were written by

Zhang Zhili, He Fang, and Ding Yao. The entire book was compiled by Ding Yao and Zhang Zhili. This book is based on the research work of the authors involved in this book and the research team led by Professor Zhang Zhili in recent years.

I would like to express my heartfelt gratitude to all the reference authors and comrades who have worked hard for the publication of this book!

With the continuous development of remote sensing image acquisition methods and processing technologies, the theory, principles, and techniques of hyperspectral image classification are also constantly evolving and improving. This book only elaborates, analyzes, and summarizes the application of graph neural network in hyperspectral image classification, hoping to provide readers with some reference and inspiration in the research of hyperspectral image classification. Due to the authors' academic level, some content research still needs to be further deepened and there are inevitably shortcomings and imperfections in the book. Readers are welcome to criticize and correct them.

Xi'an, China Yao Ding
Xi'an, China Zhili Zhang
Xi'an, China Haojie Hu
Xi'an, China Fang He
Urumqi, China Shuli Cheng
Xi'an, China Yijun Zhang

Contents

Chapter 1
Introduction

1.1 Background and Significance of the Research Project

Remote Sensing (RS) is a non-contact earth observation technology that obtains ground target information through the reflection or radiation of electromagnetic waves on ground objects [1, 2]. Military remote sensing technology can use remote sensing carriers, e.g., satellites and unmanned aerial vehicles (UAVs), to carry out high-resolution, real-time, multi-angle and multi-frequency remote sensing observation and acquisition of enemy military targets. Then, the military intelligence of enemy and battlefield conditions can be monitored, which provides intelligence support and tactical guidance for military command departments. Therefore, this technology holds a pivotal position in modern warfare, and is of great significance in improving combat efficiency and reducing war costs. In the Russia–Ukraine conflict, both sides have realized the importance of high-precision remote sensing images in obtaining the military deployment trends of the other side, and they compete to use remote sensing technology to timely understand each other's dynamics and develop corresponding strategies accordingly. At the same time, satellite remote sensing images have become the best tool for netizens and journalists to understand and analyze the conflict situation in Ukraine, enabling remote sensing technology to gain widespread attention in public opinion.

With the increase of market demand and the improvement of remote sensing technology, remote sensing technology is also developing towards high spatial-spectral resolution and high temporal resolution to carry out more accurate and faster remote sensing observation, and the reliable and timely remote sensing data is obtained. Due to the differences of imaging channels, optical remote sensing images can be divided into four categories: panchromatic image, visible image, multispectral image, and hyperspectral image. Among them, hyperspectral image data with high spectral resolution can simultaneously image the target area with tens to hundreds of continuous and fine spectral bands. In fact, different types of ground objects have different reception and radiation characteristics of electromagnetic waves, so they exhibit different

Y. Ding et al., *Graph Neural Network for Feature Extraction and Classification of Hyperspectral Remote Sensing Images*, Intelligent Perception and Information Processing, https://doi.org/10.1007/978-981-97-8009-9_1

spectral curves on the spectral bands. As shown in Fig. 1.1, after spaceborne or airborne spectral imagers image ground objects, different ground features correspond to different spectral curves in the spectral dimension. Hyperspectral image contains rich spectral information, and the ground objects that are difficult to distinguish from other optical remote sensing images under the same spatial resolution can be identified [3]. Accordingly, hyperspectral remote sensing images have been widely used in environmental monitoring, precision agriculture, geological exploration, military defense fields. For example, in environmental monitoring, hyperspectral remote sensing images can be used to monitor water quality changes in rivers, lakes, and oceans, providing important basis for pollution source tracking and environmental protection [4]; In agricultural production, the analysis of hyperspectral remote sensing data can obtain information on farmland soil nutrients, vegetation growth status, crop planting area, and yield, which helps to improve agricultural production efficiency [5]; In geological exploration, rapid identification and quantitative analysis of mineral types can be achieved by analyzing the spectral curves of different surface minerals [6]; In the field of national defense and military affairs, the analysis of hyperspectral remote sensing image data can achieve accurate identification and monitoring of enemy battlefield facilities, weapons and equipment, as well as terrain and topography, providing reliable decision-making information for the battlefield environment [7].

The hyperspectral image processing mainly includes image dimensionality reduction [8], hyperspectral unmixing [9, 10], anomaly detection [11, 12], image resolution [13], image classification [14], etc. Among them, hyperspectral classification,

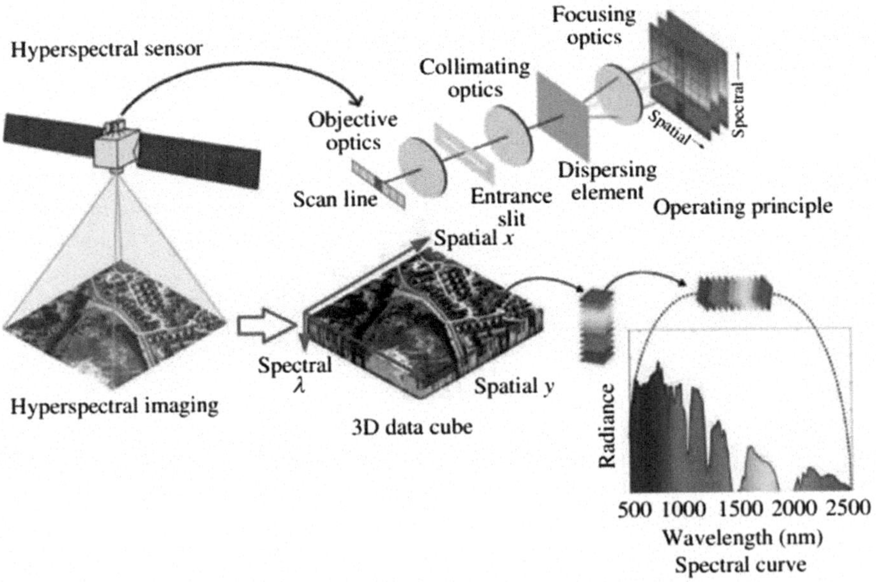

Fig. 1.1 Schematic diagram of spectrometer imaging

assigning a corresponding label to each pixel based on the spatial-spectral information of ground features, has received widespread attention through accurate recognition of ground objects as a fundamental technology of hyperspectral remote sensing image processing. It is worth noting that the improvement of spectral resolution brings challenges such as "dimensional disasters" [15], although it can provide richer ground information. The traditional hyperspectral image classification methods incur significant computational costs and cannot achieve ideal interpretation results. In addition, hyperspectral image data also faces challenges such as labeled information deficiency, isomorphism (the same type of ground objects exhibit different spectral characteristics under different conditions), and foreign object isomorphism (different types of ground objects exhibit similar spectral characteristics under specific conditions). These issues will affect the accuracy and reliability of hyperspectral image classification.

Graph Neural Network (GNN), as an emerging machine learning technology, has achieved significant results in fields such as social networks [16], bioinformatics [17], and recommendation systems [18]. Its advantage lies in its ability to naturally handle irregular and complex relational structures, such as complex but related ground features in hyperspectral remote sensing images. In this book, we aim to conduct in-depth research on the application of graph neural networks and explore their potential in hyperspectral image classification task. The GNN-based methods are expected to overcome the limitations of traditional methods and deep learning methods, and achieve higher classification accuracy and classification efficiency. This book will help to promote the development and application of hyperspectral remote sensing imaging technology, provide beneficial support for the sustainable development of the national economy, society, and ecology, and provide theoretical support and technical foundation for the updating and iteration of military remote sensing technology.

1.2 Research Status and Existing Problems

1.2.1 Traditional Hyperspectral Image Classification Methods and Existing Problems

Early hyperspectral image classification methods were mainly based on a simple combination of spectral features and conventional pattern recognition methods. However, hyperspectral image typically contains hundreds of channels, which leads to "dimensional disaster" problem for data processing and analysis [19]. To address this issue, researchers typically perform spectral dimension reduction on hyperspectral image before model training [20]. Dimension reduction methods aims to extract discriminative spectral bands or features from raw hyperspectral image data, thereby reducing image dimensionality, effectively reducing computational complexity, and improving classification accuracy. Dimension reduction methods can be divided into two categories: feature extraction methods and feature selection methods. Feature extraction methods, including principal component analysis (PCA) [21], linear discriminant analysis (LDA) [22], and independent component analysis (ICA) [23], can map high-dimensional spaces to low-dimensional spaces. Although these methods are intuitive and easy to understand, they may result in the loss of critical information during the mapping process. In contrast, the goal of feature selection methods is to improve classification accuracy by retaining the most representative spectral bands in the original hyperspectral image and discarding the bands with poor classification performance, commonly including Pap distance [24], J–M distance [24], information gain [25], and spectral angle mapping [26]. Although these methods can retain useful information from the original hyperspectral image, they usually need to be combined with search algorithms, consuming a lot of time to search for the most effective band or band combination. After dimensionality reduction processing, conventional pattern recognition methods are generally used to classify the pre-processed image, such as nearest neighbor classifiers [27], support vector machines (SVM) [28], random forests [29], decision trees [30], and sparse representations [31].

Although directly utilizing spectral features for hyperspectral image classification is relatively simple, these methods often overlook the spatial information contained in hyperspectral images (HSIs). In hyperspectral image data, both spectral information and spatial structure information contain important information about objects. Therefore, ignoring spatial structure information may loss lots of important features, which in turn affects classification performance. For this reason, many classification methods, integrating spatial and spectral information, have emerged in subsequent research. In practical applications, the following methods are usually used to extract the spatial-spectral information of HSI [32]:

(1) **Structural filtering**: Given a fixed or adaptive structural element, a series of spatial features are obtained through filtering operations. One common method is to use moment standards, such as extracting spatial information from a given area using the mean or standard deviation of adjacent pixels in the window [33]. In addition, adaptive adjustment of neighboring moments or cumulants has been widely applied in hyperspectral image classification to explore spatial homogeneity while preserving image details [34].

(2) **Mathematical morphology**: It is a powerful tool for analyzing and processing geometric structures in the spatial domain. By introducing Morphological Profile (MP), spatial structure information can be extracted from a series of open and close operations in topography for high-resolution image classification [35]. Benediktsson et al. [36] further developed the Extended Morphological Profile (EMP), which calculates MP for each dimension component after reducing the data dimension. However, this method did not fully utilize the spectral information of hyperspectral image. To reduce deficiency, Fauvel et al. [37] proposed a spectral space fusion method based on EMP and raw data by extending the concepts of MP and EMP, in which an attribute profile (AP) [38] was proposed to extract additional spatial features for HSI classification. Since then, AP and

its extensions, including extended attribute profiles [39] and extended multi attribute profiles [40], have received increasing attention in HSI classification.

(3) **Superpixel Segmentation**: Superpixel segmentation generates a set of spatially adjacent and spectral similar pixels to represent multiple local regions with different shapes and sizes. The classification performances can be improved by adopting superpixel segmentation technology to extract spectral-spatial correlated information contained in HSI [41]. For example, in reference [42], superpixel segmentation technology have been employed to extract homogeneous structural regionss, and then the graph constructed on superpixels, resulting in good classification results.

The comprehensive utilization spatial-spectral features can improve accuracy compared to HSI classification methods that rely solely on spectral features. However, these traditional methods heavily rely on manual design, which limits their feature expression capabilities, and they are difficult to meet higher-level classification requirements. In addition, the generalization ability of traditional classification methods is relatively weak, which leads that most of these methods are only applicable to specific scenarios and have poor adaptability in spectral variability of HSI [43].

1.2.2 Hyperspectral Image Classification Methods Based on Deep Learning and Existing Problems

In recent years, deep learning technology has been widely applied in many fields with the rapid development, such as image recognition, speech recognition, natural language processing, and so on. Meanwhile, many deep learning methods have been widely employed for the analysis and processing of hyperspectral images [44, 45]. Deep learning is a machine learning method based on multi-layer neural networks, which automatically adjusts the weights and biases in the neural network using backpropagation algorithms to minimize errors between training samples and labels, thereby achieving the learning and representation of high-level abstract features [46]. Compared to the shallow features obtained by traditional machine learning methods, deep learning can provide more complex and abstract structured representations of data, and the higher-level features are extracted [47, 48]. Therefore, deep learning methods have stronger robustness and invariance when dealing with adverse factors, such as changes in illumination and angle caused by sensor imaging in hyperspectral images. These advantages have made deep learning methods a research hotspot in the hyperspectral image classification field and have achieved good results. Currently, mainstream deep learning methods for hyperspectral image classification, mainly including Deep Belief Network (DBN) [49], Stacked AutoEncoder (SAE) [50], Recurrent Neural Network (RNN) [51], and Convolutional Neural Network (CNN) [52], are introduced as follows:

(1) **Deep belief network based methods**: Deep belief network is a hierarchical deep neural network that can be viewed as a concatenation of multiple restricted Boltzmann machines (RBMs), learning the feature representations of input data by unsupervised layer-by-layer structure. Peng et al. [53] first introduced deep confidence networks to extract deep spectral features, and then the high-quality labeled samples were selected as training samples for network training.

To effectively utilize the spectral and spatial information contained in HSI, Li et al. [54] proposed a DBN framework with a logistic regression layer and validated the joint utilization of spectral-spatial features can improve classification accuracy. Similarly, Sellami et al. [55] developed RBM method based on spectral-spatial information for hyperspectral image classification, in which the spectral-spatial structures were constructed through joint similarity measures, then the RBM was trained to extract useful joint spectral-spatial features for HSI classification by using DBN and logistic regression layer.

(2) **Stacked autoencoder based methods**: A stacked autoencoder is composed of multi-layer autoencoders (AE) stacked together. AE is a widely used symmetric neural network in HSI classification, which is highly favored due to its unsupervised feature learning ability. AE itself does not perform classification tasks, but provides compressed feature representations of high-dimensional HSI. Chen et al. [56] developed a stacked autoencoder to extract features of HSI and input them into a logistic regression model for classification. Zhang et al. [57] proposed an unsupervised feature extraction framework composed of a recursive autoencoder network to utilize the spectral-spatial features of HSI. To effectively address intra class variation and inter class similarity issues in HSIs, Zhou et al. [58] introduced local Fisher discriminant regularization to learn compact and discriminative features. Meanwhile, Lan et al. [59] combined the *k*-sparse denoising autoencoder and spectral–restricted spatial characteristics effectively to overcome the problem of intra class variation in spatial features.
(3) **Recurrent neural network-based methods**: Recurrent neural networks (RNNs) are neural network models that establish periodic recurrent connections between neurons at different times. Unlike traditional feedforward neural networks, the recurrent neurons in RNNs that can retain previously processed information and use them in subsequent processing. By processing the spectral information of HSI as a time series, Hang et al. [60] proposed an RNN based HSI classification framework, in which a new activation function and gating loop unit is used to determine class labels based on the sequence attributes of HSI. Zhang et al. [61] introduced an RNN framework based on local spatial sequence method, in which the Gabor filters and differential morphological contours were first adopted to extract low-level features from HSI, then these features were fused and further passed to the RNN model to extract high-level features for final classification with Softmax layer.
(4) **Convolutional neural network-based methods**: Convolutional neural networks are a type of deep learning neural networks commonly used in many fields such as image recognition and computer vision. They mainly extract

features and reduce dimensionality through multiple convolutional and pooling layers, and finally classify through fully connected layers. Convolutional operation can be seen as a filter that can detect different features in an image, such as edges, textures, etc. The pooling layer is used to reduce the spatial dimension of the image while retaining important feature information. Compared to traditional neural networks that can only utilize one-dimensional spectral vectors, CNN can effectively perceive the neighborhood information of target pixels by sliding a fixed size convolutional kernel across the entire image. Hu et al. [52] first introduced the CNN method into hyperspectral image classification, in which a one-dimensional convolutional neural network (1DCNN) was used to extract the spectral features of HSI. Zhang et al. [62] implemented a dual-channel convolutional neural network to extract spectral-spatial features for HSI classification, in which a 1DCNN was adopted to extract spectral with a layer by layer network, a 2DCNN was employ to extract hierarchical spatial features. Chen et al. [63] systematically investigated one-dimensional, two-dimensional, and three-dimensional convolutional neural networks (3DCNN), and some suggestions were proposed to guide the structural design of CNN for HSI classification.

(5) **Transformer-based methods**: Transformers can utilize self-attention (SA) mechanism to draw global dependencies in input sequences, and have been successfully applied in Natural Language Processing (NLP). Recently, some Transformer-based classifiers have been used for image processing. Sui et al. [64] proposed a Transformer called Vision Transformer (ViT) for classifying image block sequences. Carion et al. [65] proposed a new method called a detection transformer or DETR that treats object detection as a direct set prediction problem. This method directly outputs the final prediction set in parallel based on the relationship between the object and the global image context. The successful application of these Transformers has led researchers to consider whether to convert hyperspectral images into sequences for feature extraction. Hong et al. [66] proposed a new spectral transformer based on Transformer sequential perspective, which can adapt to pixel and small input blocks. Yu et al. [67] designed a multi-level spectral-spatial Transformer network (MSTNet) for hyperspectral image classification.

Recently, deep learning methods have made significant breakthroughs in hyperspectral image classification, however there are still many challenges [32, 68], namely high-dimensional spectral features and limited training samples. This book summarizes the main issues as follows:

(1) **High model complexity**: In deep neural networks (DNNs), parameter tuning and optimization are NP complete problems, and their convergence cannot be guaranteed [69–71]. Especially the applications in HSI classification, a large number of parameters need to be adjusted, and the training process is time-consuming [72]. Therefore, training DNN is considered difficulty. Designing a more concise and efficient deep network model to achieve fast and automated

hierarchical feature learning of hyperspectral remote sensing images is an urgent problem to be solved.

(2) **Limited training samples**: As aforementioned, supervised DNNs require a large amount of training data, otherwise overfitting will occur [73]. However, in practical applications, obtaining labeled samples from hyperspectral remote sensing images is very difficult. It is necessary to improve the performance of deep learning methods in small sample object recognition.

(3) **Difficulty in spatial and spectral integration learning**: Hyperspectral image confronts the problem of "same object but different spectrum, foreign object with same spectrum". So that, to achieve a higher classification accuracy, it is necessary to more effectively integrate spectral and spatial information, and extract identifiable features. However, deep learning methods for Euclidean data processing unable to effectively characterize the inter pixel relationships connected within irregular regions [20]. Therefore, more feature extraction methods need to be designed for hyperspectral image classification tasks.

1.2.3 Graph Neural Network for Hyperspectral Image Classification

To overcome the shortcomings of deep learning methods for hyperspectral image classification, Graph Neural Network (GNN) has received increasing attention. Graph neural networks is a semi-supervised framework that can perform convolution operations on non-Euclidean data [74–79]. The main reasons why graph convolutional neural networks can be well applied to hyperspectral image classification are as follows [74, 80–84]: Firstly, graph neural network methods can learn the interrelationships between nodes and have low requirements for labelled samples; In addition, graph neural networks can automatically learn and process the feature signals (spectral signals) of nodes. For hyperspectral image processing with high-dimensional spectral features, graph neural networks have natural advantages compared with convolutional neural networks; Furthermore, the combination of graph neural networks and machine learning methods can effectively distinguish the contours of landcovers and improve classification accuracy.

Graph neural networks were first proposed by Scarselli et al. [85] and are a powerful method for processing non-Euclidean data. Various graph structures, including directed graphs [86], heterogeneous graphs [87, 88], and aggregation mechanisms, have been applied to different tasks, such as convolutions [16, 89] and attention [87, 90–93]. By encoding hyperspectral images into graph structured data, graph neural networks can effectively capture and process the correlations between different object regions, while preserving the boundaries between different object categories [83]. At present, graph neural networks have been widely used in hyperspectral image classification. For example, Qin et al. [94] first proposed a semi-supervised Graph Convolutional Network (GCN) method for hyperspectral classification. This method treats each pixel in a hyperspectral image as a node, and then the GCN was adopted to process the graph constructed by nodes. However, the computation of GCN is complex. To reduce the computational cost of GCN, Chen et al. [95] proposed a miniGCN, aiming to improve the computational speed of GCN with small batch and small sample training. Sha et al. [96] introduced a graph attention mechanism in their method to replace GCN and reduce computational complexity. However, the above methods all use pixels as graph nodes, which are computationally complex and have limited applications. To address this deficiency, many superpixel-based GCN methods have emerged. For example, Wan et al. [97] introduced superpixels into GCN, which greatly reduced the number of nodes and computational complexity. Subsequently, different superpixel-based graph neural networks (GNNs) have been proposed. We proposed a multi-scale graph sample and aggregation network (GraphSAGE) for hyperspectral image classification [98], in which segmentation methods was employed to reduce the number of nodes, and spatial graph network methods was designed to reduce the computational complexity. Using graph networks for hyperspectral classification achieved high classification accuracy with small training labeled samples. However, the above methods still are semi-supervised methods, and they have not fundamentally solved the problem of dependence on labeled samples during the training process. To solve this problem, some unsupervised methods have proposed for clustering hyperspectral images. For example, Cai et al. [99] further proposed a neighborhood contrast subspace clustering (NCSC) network for large-scale hyperspectral images. In this network, the superpixels is utilized to effectively reduce computational complexity, time, and memory consumption, while greatly improving the classification accuracy of unsupervised clustering for hyperspectral images. At present, the combination of hyperspectral classification and graph convolutional neural networks is still in the exploration and development stage, with relatively few references [77, 100–102].

Graph neural network (GNN) methods have achieved significant results for hyperspectral image classification in recent years, however, existing methods still face challenges in balancing efficiency and accuracy. In addition, due to the complex and variable spectral characteristics of hyperspectral images, artificially defined graph structure data often cannot accurately reflect the complex relationships between ground objects. Therefore, this book aims to address these challenges and improve the ability of graph neural networks to extract the spectral-spatial information from hyperspectral image.

1.3 Introduction to Graph Neural Networks

In this section, a brief introduction is given to some concepts and some classic models involved in graph neural networks.

1.3.1 Concepts and Definitions Related to Graph Neural Networks

Graph: A graph is a data structure composed of a series of nodes and edges, being used to represent the relationships or interactions between objects. Given a graph represented as $\boldsymbol{G} = \{\boldsymbol{V}, \boldsymbol{E}\}$, where $\boldsymbol{V}$ represents a set of vertices or nodes, and $\boldsymbol{E}$ denotes a set of edges. In the graph, nodes represent objects, and edges represent relationships or interactions between objects. The attributes of nodes can be represented by $\boldsymbol{X} \in \boldsymbol{R}^{n\times d}$, where $n = |\boldsymbol{V}|$ denotes the number of nodes, and d is the dimension of node attributes. To indicate whether edges have directionality, graphs can be divided into directed graphs and undirected graphs, respectively.

Adjacency Matrix: Adjacency Matrix $\boldsymbol{A} \in \boldsymbol{R}^{n\times n}$ is used to represent the connection relationship between graph nodes. Let $v_i \in \boldsymbol{V}$ represent a node, and $e_{ij} = (v_i, v_j) \in \boldsymbol{E}$ denotes an edge pointing from v_j to v_i. If $e_{ij} \in \boldsymbol{E}$, then $\boldsymbol{A}_{ij} = 1$; if $e_{ij} \notin E$, then $\boldsymbol{A}_{ij} = 0$. For undirected graphs, the corresponding adjacency matrix is a symmetric matrix, while it is not necessary for a directed graph.

Degree Matrix: Degree Matrix $\boldsymbol{D} \in \boldsymbol{R}^{n\times n}$ is a diagonal matrix based on an adjacency matrix, where the elements on the diagonal are the degree of the corresponding node (the number of edges connected to that node). In a directed graph, there are two ways to define a degree matrix: out degree matrix and in degree matrix, which are used to represent the number of edges pointing to and starting from each node, respectively.

Laplacian Matrix: Laplacian matrix $\boldsymbol{L} \in \boldsymbol{R}^{n\times n}$ is a matrix used to describe the structural properties of a graph, which is the difference between the degree matrix and the adjacency matrix, and can be defined in various ways. Normally, it is defined as the symmetric normalized Laplacian matrix, namely $\boldsymbol{I} - \boldsymbol{D}^{1/2}\boldsymbol{A}\boldsymbol{D}^{-1/2}$, where $\boldsymbol{I}$ is the identity matrix.

Neighbors: Given a node v in a graph, the nodes are directly connected to it are called neighbors or neighborhood nodes, which can be defined as $\boldsymbol{N}(v) = \{u \in \boldsymbol{V} | (v, u) \in \boldsymbol{E}\}$. In the adjacency matrix, the neighbor of the nth node can be determined by searching for the position of non-zero elements in the nth row (or column) of the matrix.

Connected Component: In an undirected graph, a connected component is a subgraph, in which any two nodes are connected to each other through paths and have no connection to other nodes outside the subgraph. In a directed graph, strong connected components can be defined, where there are bidirectional paths between any two nodes. Connected components divide a graph into independent subgraphs, which may represent groups of entities with similar features or functions, and helps to understand the structure and composition of the graph.

Graph Convolution: A convolution operation performed on graph structured data. By combining the neighboring features of nodes to update their feature representations, local structural information is captured.

Aggregation Function: A function used to integrate the neighboring features of a node into the feature vector of that node in graph convolution. The commonly used aggregation functions include Mean, Sum, Max, and so on.

Graph Embedding: The process of mapping graph structured data to a low dimensional vector space. The node feature vectors learned through graph neural networks can be regarded as graph embeddings, which can capture the topological relationships and attribute information between nodes and their neighbors.

1.3.2 Classical Convolutional Graph Neural Network

Sperduti et al. first applied neural networks to directed acyclic graphs in 1997, which sparked early research interest in graph neural networks (GNNs). In 2005, Gori et al. [103] first summarized the concept of graph neural networks, which was further elaborated by Scarselli et al. [85] and Gallicchio and Micheli [104]. These early studies belong to Recursive Graph Neural Network (RecGNN), which learns the representation of target nodes by propagating neighbor information, iterating continuously until reaching a steady state. However, this process contains a high computational cost and limits its practical applications. With the successful application of convolutional neural networks (CNN), many methods have been inspired to apply the concept of convolution to graph data, which are collectively called as graph convolutional neural networks. CNNs can be divided into spectral-based methods and spatial-based methods. The spectral-based graph convolutional neural network was first proposed by Bruna et al. [105], who proposed a graph convolution method based on spectral graph theory. Subsequently, the spectral graph convolutional neural network method has been increasingly improved and expanded [68, 106]. The research on spatial graph convolutional neural networks is earlier than spectral methods. In 2009, Micheli [107] solved graph interdependence problem by constructing a non-recursive layer composite architecture, and inherited the idea of message passing from RecGNN. However, the importance of this work has not been fully recognized. Until recent years, many spatial graph convolutional neural network methods have emerged [108–110]. These studies provide a rich theoretical foundation and practical methods for the development and application of graph neural networks.

In graph neural networks, convolution operations are used for node feature extraction and representation learning. The convolution operation lies in achieving local perception on the graph, updating the feature representation of nodes through specific weight sharing and aggregation strategies [111]. To perform convolution operations on a graph, it is necessary to extend the definition of traditional convolution operations to the graph domain. The graphs convolution neural networks mainly include spectral-based methods and spatial-based methods, and we will introduce them as follows:

(1) **Spectral Graph Convolution Neural Network**

The spectral-based methods use the spectral representation of the graph for convolution operations. These methods define convolution operators in the spectral domain, and their theoretical basis is graph signal processing [112]. In spectral-based methods, a graph signal $\boldsymbol{x}$ is first converted into the spectral domain through the graph Fourier transform $\mathcal{F}$, and then convolved. After the convolution operation, the obtained signal is then converted back to the original domain through the inverse Fourier transform $\mathcal{F}^{-1}$. We can define these transformations as follows

$$\hat{\boldsymbol{x}} = \mathcal{F}(\boldsymbol{x}) = \ominus \boldsymbol{U}^T \boldsymbol{x} \tag{1.1}$$

$$\boldsymbol{x} = \mathcal{F}^{-1}(\hat{\boldsymbol{x}}) = \ominus \boldsymbol{U} \hat{\boldsymbol{x}} \tag{1.2}$$

where $\boldsymbol{U}$ is the eigenvector matrix of the normalized graph Laplace matrix $\boldsymbol{L} = \boldsymbol{I} - \boldsymbol{D}^{1/2}\boldsymbol{A}\boldsymbol{D}^{-1/2}$ ($\boldsymbol{D}$ is the degree matrix of the graph, $\boldsymbol{A}$ is the adjacency matrix of the graph), $\boldsymbol{U}^T$ denotes transposition matrix, $\hat{\boldsymbol{x}}$ is the spectral representation of the signal $\boldsymbol{x}$. The normalized graph Laplace matrix is real symmetric positive semi-definite, and it can be decomposed into $\boldsymbol{L} = \boldsymbol{U}\boldsymbol{\Lambda}\boldsymbol{U}^T$, where $\boldsymbol{\Lambda}$ is the diagonal matrix of the eigenvalues. With the convolution theorem [113], the convolution operation is defined as

$$\boldsymbol{g} * \boldsymbol{x} = \mathcal{F}^{-1}(\mathcal{F}(\boldsymbol{g}) \odot \mathcal{F}(\boldsymbol{x})) = \boldsymbol{U}\left(\boldsymbol{U}^T \boldsymbol{g} \odot \boldsymbol{U}^T \boldsymbol{x}\right) \tag{1.3}$$

where $\boldsymbol{g}$ and $\boldsymbol{x}$ are signals on the graph, $\odot$ and $*$ represent dot multiplication operation and convolution operation, respectively. The convolution operation here is performed in the spectral domain, in which the $\boldsymbol{g}$ and $\boldsymbol{x}$ are converted into the spectral domain, performing dot multiplication operations, and then converting back. In Eq. (1.3), $\boldsymbol{U}^T\boldsymbol{g}$ represents a convolutional kernel or filter. If the simplified filter is a learnable diagonal matrix $\boldsymbol{g}_w$, then the basic function of the spectral method is obtained, namely

$$\boldsymbol{g}_w * \boldsymbol{x} = \boldsymbol{U}\left(\boldsymbol{g}_w\right)\boldsymbol{U}^T \boldsymbol{x} \tag{1.4}$$

Equation (1.4) indicates that convolution operations can be simplified to matrix multiplication in the frequency domain after Fourier transform in spectral-based methods. The filter $\boldsymbol{g}_w$ determines the weight of each frequency component and can be trained to obtain the optimal weight. In addition, the eigenvalues corresponding to each feature vector actually play the role of frequency, while the filter $\boldsymbol{g}_w$ is amplitude. So, it can be seen as a spectral-based convolution method with a good efficiency and stability for dealing with large graphs.

In the following section, several typical spectral-based methods will be briefly introduced with different filters.

Spectral Network: The spectral network [105] employ a learnable diagonal matrix as a filter, namely $g_w = dig(w)$, where $w \in R^N$ is a parameter. However, this operation is computationally inefficient, and the filter is not spatially localized.

ChebNet: Defferrard et al. [68] proposed a new parameterization method called ChebNet based on Spectral Network. ChebNet adopts Chebyshev polynomials to approximate the filter g_w, and the second-order Chebyshev polynomials is used as the parameters of the filter. With this parameterization method, the filter contains the spatial localization characteristics in the frequency domain, while also having good computational efficiency and stability. Specifically, the convolution operation in ChebNet can be written as:

$$g_w * x \approx \sum_{k=0}^{K} w_k T_k(\tilde{L})x \tag{1.5}$$

where $\tilde{L} = \frac{2}{\lambda_{max}} L - I$, λ_{max} is the maximum eigenvalue of the Laplacian matrix L in the graph, the eigenvalue range of matrix $\tilde{L}$ is [−1, 1], and w_k is a learnable Chebyshev coefficient. Chebyshev polynomial is a polynomial function that can be calculated using the following recursive formula

$$T_k(x) = 2xT_{k-1}(x) - T_{k-2}(x) \tag{1.6}$$

where $T_0(x) = 1$ and $T_1(x) = 1$. A Klocalized convolutional kernels can be constructed by using K Chebyshev polynomials corresponding to K maximum eigenvalues, because these polynomials are K^{th}-order. This means that the convolution operation is only performed within the neighborhood of each node, thus limiting the influence range of the convolution kernel to local regions.

GCN: Kipf [89] proposed a simplify convolution operations method based on ChebNet, by setting K to 1 in the convolution operation to alleviate overfitting problems, and further assuming $\lambda_{max} \approx 2$, thus the convolution operation simplified to only two free parameters w_0 and w_1

$$g_w * x \approx w_0 x + w_1 (L - I)x = w_0 x - w_1 D^{-\frac{1}{2}} A D^{-\frac{1}{2}} x \tag{1.7}$$

If $w = w_0 = -w_1$, Eq. (1.7) can be expressed as follows

$$g_w * x \approx w\left(I + D^{-\frac{1}{2}} A D^{-\frac{1}{2}} x\right) \tag{1.8}$$

To solve the problem of gradient vanishing or exploding in Eq. (1.9), a renormalization technique is further introduced, namely $I + D^{-1/2} A D^{-1/2} \to \tilde{D}^{-1/2} \tilde{A} \tilde{D}^{-1/2}$, where $\tilde{A} = A + I$, $\tilde{D}_{ii} = \sum_j \tilde{A}_{ij}$. Finally, the matrix representation of GCN is defined as

$$H = \tilde{D}^{-1/2} \tilde{A} \tilde{D}^{-1/2} XW \tag{1.9}$$

where $X \in R^{N \times F}$ is the input feature matrix, N represents the number of nodes, F denotes the feature dimension of each node. $W \in R^{F \times F}$ is the parameter matrix, F represents the dimension of the output feature, $H \in R^{N \times F}$ is the output feature matrix.

(2) **Spatial Graph Convolution Neural Network**

Spatial graph convolution neural network directly defines convolution operations based on graph topology, in which how to define convolution operations with different size neighborhoods while maintaining local invariance is the biggest challenge. In traditional convolution neural networks, convolutional operations are performed on regular structures such as images. If an image is viewed as a special form of graph structure in which each pixel represents a node, each pixel directly establishes a connection with its neighboring pixels, as shown in Fig. 1.2a. By applying the 3 × 3 convolution kernel on the image, we can obtain the convolution result of the target region by the weighted average of the pixel values between central node and its neighbors. The spatial graph convolution neural network is similar to the convolution operation of traditional CNNs on images, defining graph convolution based on the spatial relationships of nodes. Similarly, spatial graph convolution neural network convolves the representation of the central node with the representation of its neighbors to derive the updated representation of the central node, as shown in Fig. 1.2b. It can be seen that spatial graph convolution neural network shares the same idea of information propagation as CNN, which operations essentially propagate node information along edges. However, due to the different degrees of nodes in irregular structured data (i.e., the number of edges connected to nodes), spatial graph convolution neural network needs to address how to define convolutional operations with different sizes of neighbors while maintaining local invariance.

GraphSage: In graph data, the number of neighbors of a node may vary from one to one thousand or even more, taking all samples of the node's neighborhood for

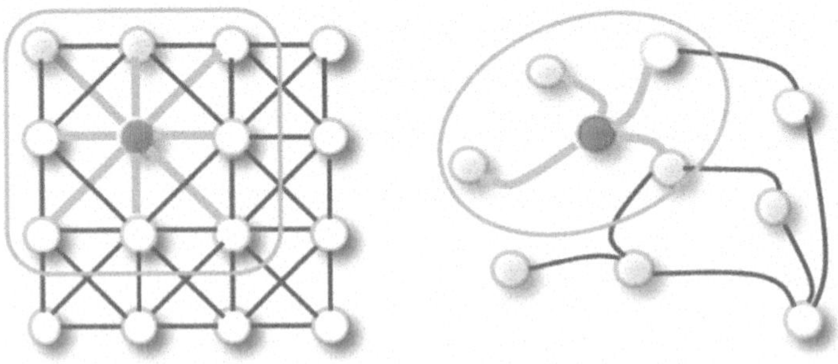

(a) 2D Convolution (b) Graph Convolution

Fig. 1.2 Schematic diagram of convolution and graph convolution principles

convolution is inefficient. GraphSage [16] employ sampling to obtain a fixed number of neighbors for each node, which can be presented as follows

$$h_v^{(k)} = \sigma\left(W(k)f_k\left(h_v^{(k1)}, \{h_u^{(k1)}, \forall u \in \mathrm{N}_{(v)}\}\right)\right) \tag{1.10}$$

where $h_v^{(0)} = x_v$, $f_k(\cdot)$ is an aggregation function that converts the information of this neighbor subset into a new representation of the central node. $\mathrm{N}_{(v)}$ is a randomly sampled sample of the nearest neighbor of a node. The order of neighboring nodes in a graph can be arbitrary, so the aggregation function should maintain the permutation of the order of neighboring nodes unchanged. That is to say, if the order of any set of neighboring nodes is permuted, the output of the aggregation function should remain unchanged, which is called local invariance and is an important property for designing and calculating graph aggregation functions. Commonly used graph aggregation functions include mean, sum, or maximum functions.

Graph Attention Network (GAT) [114]: In many applications, the relationships between nodes may be complex, and their relationships may change over time or tasks. At this point, it is unreasonable to assume that the contribution weight of neighboring nodes to the central node is fixed. Unlike graph neural networks such as GraphSage, assuming each neighboring node has the same contribution weight to the central node. GAT adopts an attention mechanism that allows the central node to dynamically calculate the contribution weight of each neighboring node. With this way, the central node can better extract the information of neighboring nodes, thereby improving the performance of the entire graph neural network. The graph convolution operation in GAT can be represented as

$$h_v^{(k)} = \sigma\left(\sum_{u \in N(v) \cup v} \alpha_{vu}^{(k)} \mathbf{W}^{(k)} h_u^{(k-1)}\right) \tag{1.11}$$

where $h_v^{(0)} = x_v$, $W^{(k)}$ and σ are the weight matrix and the activation function, respectively. α_{uv} denotes the attention weight coefficient between node v and its nearest neighbor u, namely

$$\alpha_{vu}^{(k)} = softmax\left(g\left(\boldsymbol{a}^T\left(\mathbf{W}^{(k)} h_v^{(k-1)} \| \mathbf{W}^{(k)} h_u^{(k-1)}\right)\right)\right) \tag{1.12}$$

where $g(\cdot)$ is the activation function of LeakyReLU, $\boldsymbol{a}$ is a learnable parameter vector. Softmax function is used to ensure that the sum of attention weights for all nearest neighbors of a node is 1. GAT further adopts a multi-head attention mechanism to improve the expression ability. Specifically, GAT uses a multi-head attention mechanism to calculate the contribution weight of each neighboring node to the central node. Each attention head calculates the weight of a set of neighboring nodes, and then combines these weights to obtain the final contribution weight of the neighboring nodes. In this way, GAT can dynamically learn the relationships between each node based on the structure of the graph, thereby improving the performance of the entire graph neural network.

Message Passing Neural Network (MPNN): MPNN [115] is a universal framework based on spatial graph convolutional neural networks, which defines graph convolution as a message passing process directly transferring information from one node to another along the edge. MPNN runs K-step message passing iterations to achieve long-distance information dissemination. MPNN propagates information based on message passing mechanisms, typically including message aggregation and update stages. The hidden presentation h_v^l of each node is updated based on message m_v^{l+1}, namely

$$m_v^{l+1} = \sum_{w \in N(v)} M\left(h_v^l, h_w^l, e_{vw}\right) \tag{1.13}$$

$$h_v^{l+1} = U\left(h_v^l, m_v^{l+1}\right) \tag{1.14}$$

where m_v^{l+1} is the information received by node v in $l+1$ time step, $N(v)$ is the neighborhood set node v, h_v^l is the eigenvector of node v in t time step, e_{vw} is the edge feature of nodes v and w, and $M(\bullet)$ is the message functions, including a mean aggregator [89], a maximum pooling aggregator [116, 117], attention aggregator [118], or LSTM aggregator [119]. $U(\bullet)$ is the node update function (including a multi-layer perceptron [120], a gated network [80], and so on.), which takes the original node state h_v^l and information m_v^{l+1} as inputs to obtain the new node state h_v^{l+1}. The Eq. (1.13) is the same as the update function of RNN. The classic GCN model can also be seen as a special case of MPNN, where the message function is $M(h_v^l, h_w^l) = c_{vw}h_w^l$ ($c_{vw} = \left(deg(v)deg(w)^{-1/2}\right)A_{vw}$), and the node update function is $U(h_v^l, m_v^{l+1}) = ReLU(W^l m_v^{l+1} + 1)$.

The spectral method has a theoretical foundation in graph signal processing. By utilizing the eigenvalues and eigenvectors of graphs to design new graph filters, convolution operations will contain good interpretability. However, spectral-based methods are more inefficient than spatial-based methods because they either require feature vector calculations or require simultaneous processing of the entire graph. The spatial-based methods perform convolution directly in the spatial domain through information transfer, which can simultaneously process nodes in parallel, making them more suitable for large-scale graph representation. In addition, spectral spectral-based methods based on graph Fourier transform have poor generalization ability when dealing with new graphs, as they assume that the graph is fixed and any disturbance to the graph will result in changes in the eigenvectors. On the contrary, spatial-based methods perform local graph convolution on each node, making it easier to share weights between different positions and structures, and more flexible. In summary, due to the efficiency, generality, and flexibility, the spatial-based method is more popular than the spectral-based method.

1.4 Evaluation Indicators

In this book, six indicators are adopted to evaluate the performance of the investigated methods, mainly including Overall Accuracy (OA), Personal Accuracy (PA), Average Accuracy (AA), Kappa coefficient (κ), Normalized Mutual Information (NMI), and Adjusted Rand Index (ARI). Firstly, we need to introduce the error matrix, which can be expressed as

$$M = \begin{bmatrix} m_{11} & m_{12} & \cdots & m_{1C} \\ m_{21} & m_{22} & \cdots & m_{2C} \\ \vdots & \vdots & \ddots & \vdots \\ m_{C1} & m_{C2} & \cdots & m_{CC} \end{bmatrix} \tag{1.15}$$

where m_{ij} represents the number of samples recognized as j classes in the i class, and C is the number of classes in the image.

Overall accuracy (OA): Refers to the ratio of the number of correctly classified samples to the total number of samples in the image, which can be expressed as

$$OA = \sum_{i=1}^{C} m_{ii}/N \tag{1.16}$$

where N represents the number of samples in the image.

Precision per class (PA): Refers to the ratio of the number of correctly classified samples in each class of the calculated image to the total number of samples in each class, which can be expressed as

$$PA = m_{ii} \big/ N \tag{1.17}$$

where N_i represents the number of samples in i class in the image.

Average Accuracy (AA): Refers to the average accuracy of each category in the image, which can be expressed as

$$AA = \sum_{i=1}^{C} \frac{m_{ii}}{N_i}/C \tag{1.18}$$

Kappa coefficient (κ): Refers to the similarity between the image classification results and the ground-truth image, namely

$$Kappa = N\sum_{i=1}^{C} m_{ii} - \sum_{i=1}^{C} m_{i+}m_{+i} \Big/ N^2 - \sum_{i=1}^{C} m_{i+}m_{+i} \tag{1.19}$$

where m_{i+} and m_{+i} represent the row and column of Eq. (1.15).

1.5 Research Content and Chapter Arrangement

In this book, the graph learning methods are designed to solve the problems and challenges faced by hyperspectral image classification and improve classification accuracy, focusing on model construction, improving graph information dissemination methods, and improving composition quality. Multiple graph learning based hyperspectral image classification methods are proposed. The research content and their interrelationships are shown in Fig. 1.3, which can be summarized into the following aspects:

This Chapter. Introduction

This chapter mainly introduces the research background of this book, explains the practical significance of hyperspectral image classification, and reviews the research status and existing problems of hyperspectral image classification. To facilitate the understanding of this book, the basic knowledge of graph networks and the evaluation indicators used in the experiment were introduced. Finally, the research content and chapter arrangement of this book were specifically presented.

Chapter 2. Graph Sample and Aggregate-Attention Network for Hyperspectral Image Classification

Graph Convolutional Network (GCN) has shown potential in hyperspectral image (HSI) classification. However, GCN is a transductive learning method, which is difficult to aggregate the new node. And the available GCN-based methods fail to understand the global and contextual information of the graph. To address this deficiency, a novel semisupervised network based on graph sample and aggregate-attention (SAGE-A) for HSIs classification is proposed. Different from the graph convolution neural network (GCN) based method, SAGE-A adopts a multi-level graph sample and aggregate (graphSAGE) network, as it can flexibly aggregate the new neighbor node among arbitrarily structured non-Euclidean data and capture long-range contextual relations. Inspired by the convolution neural network (CNN) self-attention mechanism, the proposed network employs the graph attention mechanism to characterize the importance among spatially neighboring regions, so the deep contextual and global information of the graph can be learned automatically by focusing on important spatial targets.

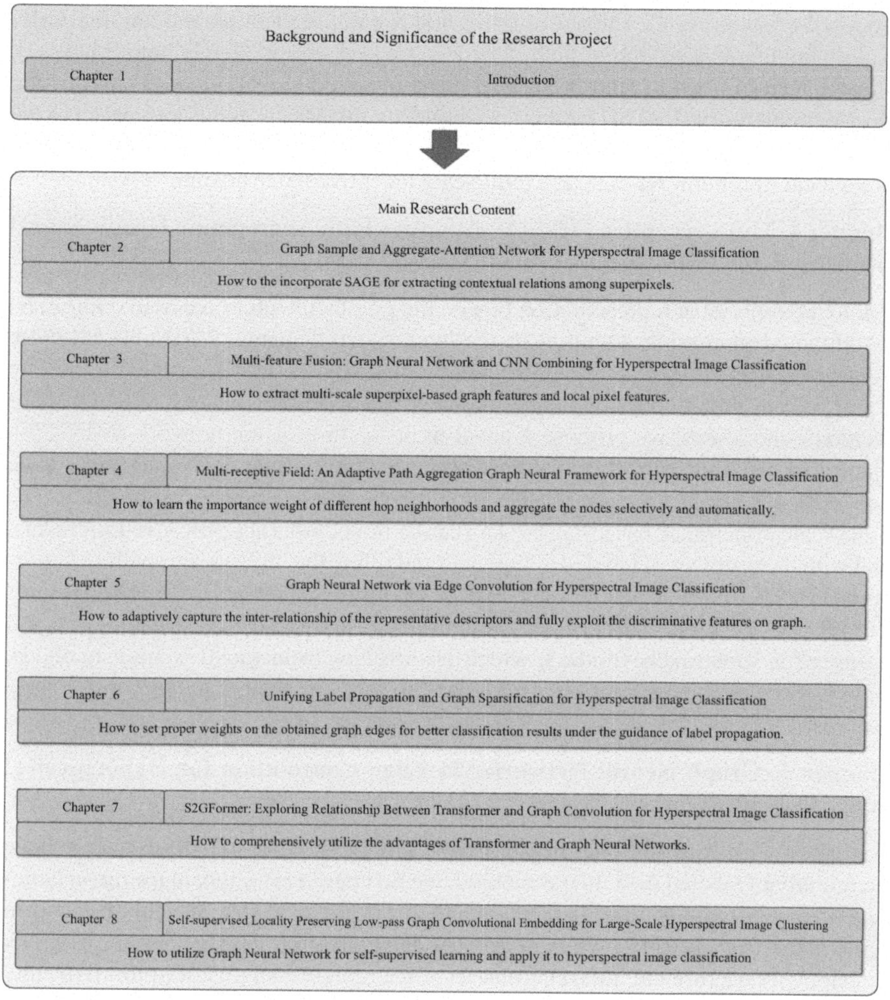

Fig. 1.3 The main research content of this book

Chapter 3. Multi-feature Fusion: Graph Neural Network and CNN Combining for Hyperspectral Image Classification

The applications of graph convolutional networks (GCNs) in hyperspectral image (HSI) classification have attracted much attention. However, hyperspectral classification faces problems such as complex noise effects, spectral variability, labelled training sample deficiency, and high spectral mixing between materials. Furthermore, the available GCN-based methods are computationally complex and cannot automatically adjust aggregate paths. To mitigate these issues, we propose a novel multiadaptive receptive field-based graph neural framework (MARP) for HSI classification. In our method, an adaptive receptive path aggregation (ARP) mechanism is proposed to suppress the impact of noise nodes on classification and automatically explore an adaptive receptive field, where a graph attention (GAT) neural network is introduced to learn the importance of different-sized neighbourhoods and a long short-term memory (LSTM) method is adopted to update the nodes and preserve the local convolutional features of the nodes. To address the problem that ARP may fall into a local optimum, we design a multiscale receptive mechanism.

Chapter 4. Multi-receptive Field: An Adaptive Path Aggregation Graph Neural Framework for Hyperspectral Image Classification

Due to its impressive representation power, the graph convolutional network (GCN) has attracted increasing attention in the hyperspectral image (HSI) classification. However, the most of available GCN-based methods for HSI classification utilize superpixels as graph nodes, which ignore the pixel-wise spectral-spatial features. To overcome the issues, we propose a novel multi-feature fusion network (MFGCN), where two different convolutional networks, i.e., multi-scale GCN and multi-scale convolutional neural network (CNN), are utilized in two branches, separately. The multi-scale superpixel-based GCN can reduce the computing power requirements, deal with the problem of labeled deficiency, and refine the multi-scale spatial features from HSI. The multi-scale CNN can extract the multi-scale pixel-wise local features for HSI classification. Furthermore, we introduced a 1D CNN to extract the spectral features for superpixels (nodes), which is different from most existing methods. Finally, a concatenate operation is employed to fuse the complementary multi-scale features.

Chapter 5. Graph Neural Network Via Edge Convolution for Hyperspectral Image Classification

Compared with convolutional neural network (CNN), GNN can effectively relieve the scarcity of labeled data. In our method, we first perform feature learning on large-scale irregular regions through GNN and then extract local spatial–spectral features at the pixel level. Besides, we incorporate edge convolution (EdgeConv) into GNN to adaptively capture the interrelationship of the representative descriptors and fully exploit the discriminative features on graph.

Chapter 6. Unifying Label Propagation and Graph Sparsification for Hyperspectral Image Classification

The existing GCN-based models for HSIC propagate and aggregate information through the GCN network based on the graph, which is constructed according to spatial location or spectral similarity. However, the constructed graph may not be ideal for the downstream classification task due to the variety of spectral characteristics. In this letter, a fully connected graph is adaptively constructed to make full use of local spatial information and global spectral information. Besides, we apply a neural sparsification technique to remove potentially task-irrelevant edges in the case of misleading message propagation. Furthermore, label propagation (LP) serves as regularization to assist the graph network in learning proper edge weights that lead to improved classification performance.

Chapter 7. S²GFormer: Exploring Relationship Between Transformer and Graph Convolution for Hyperspectral Image Classification

Recently both graph convolutional neural networks (GCNs) and Transformers have shown promising progress in hyperspectral image (HSI) classification. Transformer-based methods have a great ability to model non-local interactions among spectral and spatial information, whereas GCNs tend to do well in exploiting neighborhood vertex interactions based on their unique aggregation mechanism. In this paper, we study how to comprehensively utilize the advantages of transformer and graph convolution by combing the two structures into a unified Transformer (Graphormer) to model both global and local interactions for HSI classification, and spatial-spectral features enhanced Graphormer framework (S^2GFormer) is proposed. Specifically, a Follow Patch mechanism is first proposed to transform the pixel in HSI to patches while preserving the local spatial features and reducing the computation cost. Moreover, a patch-wise spectral embedding block is designed to extract the spectral features of the patch, in which a neighborhood convolution is developed to extract comprehensive spectral information of HSI. Finally, a multi-layer Graphormer encoder module is proposed to extract the representative spatial-spectral features from the patch for HSI classification. In our network, the three aforementioned parts are jointly integrated into a unified end-to-end network, and each component benefits the other.

Chapter 8. Self-Supervised Locality Preserving Low-Pass Graph Convolutional Embedding for Large-Scale Hyperspectral Image Clustering

Hyperspectral image (HSI) clustering is extremally a fundamental but challenging task, due to prior knowledge deficiency, large spectral variability, and high-dimension of HSI. Deep clustering methods has achieved remarkable success and has attracted increasing attentions in the unsupervised hyperspectral image classification (HSIC). However, the poor robustness, adaptability, and feature presentation limit their practical application to complex large-scale HSI datasets. Thus, this paper introduces a novel self-supervised locality preserving low-pass graph convolutional embedding method (L2GCC) for large-scale hyperspectral image clustering. Specifically, a spectral-spatial transformation HSI pre-processing mechanism is introduced to learn superpixel-level spectral-spatial features from HSI and reduce the number of graph nodes for subsequent network processing. In addition, locality preserving low-pass graph convolutional embedding autoencoder is proposed, in which the low-pass graph convolution and layer-wise graph attention are designed to extract the smoother features and preserve layer-wise locality features, respectively. Finally, we develop a self-training strategy, in which self-training clustering objective employs soft labels to supervise the clustering process and obtain appropriate hidden representations for node clustering.

References

1. SambhuNath, An introduction to remote sensing. Koros (2014)
2. A.B. Brink, C. Schmidt, Z. Szantoi, Introduction to remote sensing for conservation practitioners (2018)
3. P. Mchaffie, S. Hwang, C. Follett, Introduction to remote sensing and GIS (2018)
4. S. Somvanshi, M. Kumari, An introduction to remote sensing and its applications (2014)
5. Y. Dua, V. Kumar, R.S. Singh, Comprehensive review of hyperspectral image compression algorithms. Opt. Eng. **59**, 9 (2020)
6. A. Kaul, S. Raina, Support vectormachine versus convolutional neural network for hyperspectral image classification: a systematic review. Concurr. Comput. Pract. Exper. (2022)
7. S.P. Sabale, A.C.R. Jadhav, Supervised, unsupervised, and semisupervised classification methods for hyperspectral image classification: a review. Int. J. Sci. Res. **3**, 2319–7064 (2014)
8. L.M. Bruce, C.H. Koger, J. Li, Dimensionality reduction of hyperspectral data using discrete wavelet transform feature extraction. IEEE Trans. Geosci. Remote Sens. **40**(10), 2331–2338 (2002)
9. J.M. Bioucas-Dias, A. Plaza, N. Dobigeon, M. Parente, Q. Du, P. Gader, J. Chanussot, Hyperspectral unmixing overview: geometrical, statistical, and sparse regression-based approaches. IEEE J. Select. Top. Appl. Earth Observ. Remote Sens. **5**(2), 354–379 (2012)
10. R. Heylen, M. Parente, P. Gader, A review of nonlinear hyperspectral unmixing methods. IEEE J. Select. Top. Appl. Earth Observ. Remote Sens. **7**(6), 1844–1868 (2014)
11. J. Liu, Z. Hou, W. Li, R. Tao, D. Orlando, H. Li, Multipixel anomaly detection with unknown patterns for hyperspectral imagery. IEEE Trans. Neural Netw. Learn. Syst. **33**(10), 5557–5567 (2021)
12. W. Rao, Y. Qu, L. Gao, X. Sun, Y. Wu, B. Zhang, Transferable network with Siamese architecture for anomaly detection in hyperspectral images. Int. J. Appl. Earth Observ. Geoinform. **106**, 102669 (2022)
13. R. Ran, L.-J. Deng, T.-X. Jiang, J.-F. Hu, J. Chanussot, G. Vivone, GuidedNet: a general CNN fusion framework via high-resolution guidance for hyperspectral image super-resolution. IEEE Trans. Cybern. (2023)
14. Y. Dong, Q. Liu, B. Du, L. Zhang, Weighted feature fusion of convolutional neural network and graph attention network for hyperspectral image classification. IEEE Trans. Image Process. **31**, 1559–1572 (2022)
15. J.M. Bioucas-Dias, A. Plaza, G. Camps-Valls, P. Scheunders, N. Nasrabadi, J. Chanussot, Hyperspectral remote sensing data analysis and future challenges. IEEE Geosci. Remote Sens. Mag. **1**, 6–36 (2013)
16. W. Hamilton, Z. Ying, J. Leskovec, Inductive representation learning on large graphs. Adv. Neural Inform. Process. Syst. **30**, 11542 (2017)
17. A. Fout, J. Byrd, B. Shariat, A. Ben-Hur, Protein interface prediction using graph convolutional networks. Adv. Neural Inform. Process. Syst. **30**, 354 (2017)
18. X. Wang, X. He, Y. Cao, M. Liu, T.-S. Chua, Kgat: knowledge graph attention network for recommendation, pp. 950–958
19. A. Sellami, S. Tabbone, Deep neural networks-based relevant latent representation learning for hyperspectral image classification. Pattern Recogn. **121**, 108224 (2022)
20. J. Wang, X. Tan, J. Lai, J. Li, ASPCNet: deep adaptive spatial pattern capsule network for hyperspectral image classification. Neurocomputing **486**, 47–60 (2022)
21. L. Wei, Locality-preserving dimensionality reduction and classification for hyperspectral image analysis. IEEE Trans. Geosci. Remote Sens. **50**(4), 1185–1198 (2012)
22. T.V. Bandos, L. Bruzzone, G. Camps-Valls, Classification of hyperspectral images with regularized linear discriminant analysis. IEEE Trans. Geosci. Remote Sens. **47**(3), 862–873 (2009)
23. A. Agarwal, T. El-Ghazawi, H. El-Askary, J. Le-Moigne, Efficient hierarchical-PCA dimension reduction for hyperspectral imagery, in *Proceedings of the Signal Processing and Information Technology* (2007)

24. N. Keshava, Distance metrics and band selection in hyperspectral processing with applications to material identification and spectral libraries. IEEE Trans. Geosci. Remote Sens. **42**(7), 1552–1565 (2004)
25. L. Bruzzone, F. Roli, An extension of the Jeffreys-Matusita distance to multiclass cases for feature selection. IEEE Trans. Geosci. Remote Sens. **33**(6), 1318–1321 (1995)
26. T. Kailath, The divergence and bhattacharyya distance measures in signal selection. IEEE Trans. Commun. Technol. **15**(1), 52–60 (1967)
27. M.A. Hossain, M. Pickering, X. Jia, Unsupervised feature extraction based on a mutual information measure for hyperspectral image classification, in *Proceedings of the IEEE* (2011), pp. 1720–1723
28. F. Melgani, L. Bruzzone, Classification of hyperspectral remote sensing images with support vector machines. IEEE Trans. Geosci. Remote Sens. **42**(8), 1778–1790 (2004)
29. J. Ham, Y. Chen, M.M. Crawford, J. Ghosh, Investigation of the random forest framework for classification of hyperspectral data. IEEE Trans. Geosci. Remote Sens. **43**(3), 492–501 (2005)
30. S. Delalieux, B. Somers, B. Haest, T. Spanhove, J.V. Borre, C. Mücher, Heathland conservation status mapping through integration of hyperspectral mixture analysis and decision tree classifiers. Remote Sens. Environ. **126**, 222–231 (2012)
31. Y. Chen, N.M. Nasrabadi, T.D. Tran, Hyperspectral image classification using dictionary-based sparse representation. IEEE Trans. Geosci. Remote Sens. **49**(10), 3973–3985 (2011)
32. S. Wan, C. Gong, P. Zhong, S. Pan, G. Li, J. Yang, Hyperspectral image classification with context-aware dynamic graph convolutional network. IEEE Trans. Geosci. Remote Sens. **59**(1), 597–612 (2020)
33. L. He, J. Li, C. Liu, S. Li, Recent advances on spectral-spatial hyperspectral image classification: an overview and new guidelines. IEEE Trans. Geosci. Remote Sens. **99**, 1–19 (2017)
34. H. Xin, L. Zhang, An adaptive mean-shift analysis approach for object extraction and classification from urban hyperspectral imagery. IEEE Trans. Geosci. Remote Sens. **46**(12), 4173–4185 (2008)
35. M. Pesaresi, J.A. Benediktsson, A new approach for the morphological segmentation of high-resolution satellite imagery. IEEE Trans. Geosci. Remote Sens. **39**(2), 309–320 (2002)
36. J.A. Benediktsson, J.A. Palmason, J.R. Sveinsson, Classification of hyperspectral data from urban areas based on extended morphological profiles. IEEE Trans. Geosci. Remote Sens. **43**(3), 480–491 (2005)
37. M. Fauvel, J. Chanussot, N.A. Benediktsson, A spatial-spectral kernel-based approach for the classification of remote-sensing images. Pattern Recogn. **45**, 381–392 (2012)
38. M.M. Dalla, J.A. Benediktsson, B. Waske, B. Lorenzo, Morphological attribute profiles for the analysis of very high resolution images. IEEE Trans. Geosci. Remote Sens. **48**(10), 3747–3762 (2010)
39. M.D. Mura, A. Villa, J.A. Benediktsson, J. Chanussot, L. Bruzzone, Classification of hyperspectral images by using extended morphological attribute profiles and independent component analysis. IEEE Geosci. Remote Sens. Lett. **8**(3), 542–546 (2011)
40. M. Pedergnana, P.R. Marpu, M.D. Mura, J.A. Benediktsson, L. Bruzzone, A novel technique for optimal feature selection in attribute profiles based on genetic algorithms. IEEE Trans. Geosci. Remote Sens. **51**(6), 3514–3528 (2013)
41. S. Subudhi, R. Narayan, P.K. Biswal, F. Dell'Acqua, A survey on superpixel segmentation as a pre-processing step in hyperspectral image analysis. IEEE J. Select. Top. Appl. Earth Observ. Remote Sens. (2021)
42. B. Cui, X. Xie, X. Ma, G. Ren, Y. Ma, Superpixel-based extended random walker for hyperspectral image classification. IEEE Trans. Geosci. Remote Sens. **12**, 1–11 (2018)
43. Y. Xu, B. Du, F. Zhang, L. Zhang, Hyperspectral image classification via a random patches network. ISPRS J. Photogramm. Remote Sens. **142**, 344–357 (2018)
44. H. Petersson, D. Gustafsson, D. Bergström, Hyperspectral image analysis using deep learning: a review. Tools Appl. (2017)

45. M. Ahmad, A.M. Khan, M. Mazzara, S. Distefano, M.S. Sarfraz, A fast and compact 3-D CNN for hyperspectral image classification. IEEE Geosci. Remote Sens. Lett. **12**, 1–5 (2021)
46. G.E. Hinton, R.R. Salakhutdinov, Reducing the dimensionality of data with neural networks. Science **313**(5786), 504–507 (2006)
47. Z. Qin, L. Ni, Z. Tong, W. Qian, Deep learning based feature selection for remote sensing scene classification. IEEE Geosci. Remote Sens. Lett. **12**(11), 1–5 (2015)
48. H. Fan, X. Gui-Song, H. Jingwen, Z. Liangpei, Transferring deep convolutional neural networks for the scene classification of high-resolution remote sensing imagery. Remote Sens. **7**(11), 14680–14707 (2015)
49. G.E. Hinton, S. Osindero, Y.W. Teh, *A Fast Learning Algorithm for Deep Belief Nets* (MIT Press, 2006)
50. J. Xu, X. Lei, R. Hang, J. Wu, *Stacked Sparse Autoencoder (SSAE) for Nuclei Detection on Breast Cancer Histopathology Images* (IEEE, 2014)
51. R.J. Williams, D. Zipser, A learning algorithm for continually running fully recurrent neural networks. Neural Comput. **1**, 2 (1998)
52. W. Hu, Y.Y. Huang, L. Wei, F. Zhang, H.C. Li, Deep convolutional neural networks for hyperspectral image classification. J. Sens. **2015**, 5487 (2015)
53. L. Peng, Z. Hui, K.B. Eom, Active deep learning for classification of hyperspectral images. IEEE J. Select. Top. Appl. Earth Observ. Remote Sens. **10**(2), 712–724 (2017)
54. S. Li, W. Song, L. Fang, Y. Chen, P. Ghamisi, J.A. Benediktsson, Deep learning for hyperspectral image classification: an overview. IEEE Trans. Geosci. Remote Sens. **57**(9), 6690–6709 (2019)
55. A. Sellami, I. Farah, Spectra-spatial graph-based deep restricted boltzmann networks for hyperspectral image classification, in *Proceedings of the 2019 PhotonIcs and Electromagnetics Research Symposium-Spring (PIERS-Spring)* (2019), pp. 1055–1062
56. Y. Chen, Z. Lin, X. Zhao, G. Wang, Y. Gu, Deep learning-based classification of hyperspectral data. IEEE J. Select. Top. Appl. Earth Observ. Remote Sens. **7**(6), 2094–2107 (2014)
57. X. Zhang, Y. Liang, L. Chen, H. Ning, L. Jiao, H. Zhou, Recursive autoencoders-based unsupervised feature learning for hyperspectral image classification. IEEE Geosci. Remote Sens. Lett. **99**, 1–5 (2017)
58. P. Zhou, J. Han, G. Cheng, B. Zhang, "Learning compact and discriminative stacked autoencoder for hyperspectral image classification. IEEE Trans. Geosci. Remote Sens. **57**(7), 4823–4833 (2019)
59. R. Lan, Z. Li, Z. Liu, T. Gu, X. Luo, Hyperspectral image classification using k-sparse denoising autoencoder and spectral–restricted spatial characteristics. Appl. Soft Comput. **74**, 693–708 (2019)
60. R. Hang, Q. Liu, D. Hong, P. Ghamisi, Cascaded recurrent neural networks for hyperspectral image classification. IEEE Trans. Geosci. Remote Sens. **57**(8), 5384–5394 (2019)
61. X. Zhang, Y. Sun, J. Kai, L. Chen, L. Jiao, H. Zhou, Spatial sequential recurrent neural network for hyperspectral image classification. IEEE J. Select. Top. Appl. Earth Observ. Remote Sens. **12**, 1–15 (2018)
62. H. Zhang, Y. Li, Y. Zhang, Q. Shen, Spectral-spatial classification of hyperspectral imagery using a dual-channel convolutional neural network. Remote Sens. Lett. **8**(5), 438–447 (2017)
63. Y. Chen, H. Jiang, C. Li, X. Jia, P. Ghamisi, Deep feature extraction and classification of hyperspectral images based on convolutional neural networks. IEEE Trans. Geosci. Remote Sens. **54**(10), 6232–6251 (2016)
64. B. Sui, T. Jiang, Z. Zhang, X. Pan, ECGAN: an improved conditional generative adversarial network with edge detection to augment limited training data for the classification of remote sensing images with high spatial resolution. IEEE J. Select. Top. Appl. Earth Observ. Remote Sens. **14**, 1311–1325 (2020)
65. N. Carion, F. Massa, G. Synnaeve, N. Usunier, A. Kirillov, S. Zagoruyko, *End-to-End Object Detection with Transformers*. pp. 213–229
66. D. Hong, Z. Han, J. Yao, L. Gao, B. Zhang, A. Plaza, J. Chanussot, SpectralFormer: rethinking hyperspectral image classification with transformers. IEEE Trans. Geosci. Remote Sens. **60**, 1–15 (2021)

67. H.Y. Yu, Z. Xu, K. Zheng, D.F. Hong, H. Yang, M.P. Song, MSTNet: a multilevel spectral-spatial transformer network for hyperspectral image classification. IEEE Trans. Geosci. Remote Sens. **60**, 1152 (2022)
68. M. Defferrard, X. Bresson, P. Vandergheynst, Convolutional neural networks on graphs with fast localized spectral filtering. Adv. Neural Inform. Process. Syst. **29**, 956 (2016)
69. Q. Nguyen, M. Hein, *Optimization Landscape and Expressivity of Deep CNNs* (2018)
70. M. Ahmad, S. Shabbir, R.A. Raza, M. Mazzara, A.M. Khan, Artifacts of different dimension reduction methods on hybrid CNN feature hierarchy for hyperspectral image classification. Optik Int. J. Light Electr. Opt. **1**, 167757 (2021)
71. M. Ahmad, M. Mazzara, S. Distefano, 3D/2D regularized CNN feature hierarchy for hyperspectral image classification. Remote Sens. (2021)
72. C. Si, Y. Wang, *Convolutional Neural Network and Convex Optimization*
73. D. Erhan, Y. Bengio, A. Courville, P.A. Manzagol, P. Vincent, S. Bengio, Why does unsupervised pre-training help deep learning? J. Mach. Learn. Res. **11**(3), 625–660 (2010)
74. F. Luo, L. Zhang, B. Du, L. Zhang, Dimensionality reduction with enhanced hybrid-graph discriminant learning for hyperspectral image classification. IEEE Trans. Geosci. Remote Sens. **58**(8), 5336–5353 (2020)
75. P. Yang, L. Tong, B. Qian, Z. Gao, J. Yu, C. Xiao, Hyperspectral image classification with spectral and spatial graph using inductive representation learning network. IEEE J. Select. Top. Appl. Earth Observ. Remote Sens. **14**, 791–800 (2020)
76. L. Mou, X. Lu, X. Li, X.X. Zhu, Nonlocal graph convolutional networks for hyperspectral image classification. IEEE Trans. Geosci. Remote Sens. **58**(12), 8246–8257 (2020)
77. G. Shi, H. Huang, Z. Li, Y. Duan, Multi-manifold locality graph preserving analysis for hyperspectral image classification. Neurocomputing **388**, 45–59 (2020)
78. P. Sellars, A.I. Aviles-Rivero, C.-B. Schönlieb, Superpixel contracted graph-based learning for hyperspectral image classification. IEEE Trans. Geosci. Remote Sens. **58**(6), 4180–4193 (2020)
79. M. Sharma, M. Biswas, Classification of hyperspectral remote sensing image via rotation-invariant local binary pattern-based weighted generalized closest neighbor. J. Supercomput. **77**(6), 5528–5561 (2021)
80. S. Wan, C. Gong, S. Pan, J. Yang, J. Yang, *Multi-Level Graph Convolutional Network with Automatic Graph Learning for Hyperspectral Image Classification* (2020)
81. S. Jia, X. Deng, M. Xu, J. Zhou, X. Jia, Superpixel-level weighted label propagation for hyperspectral image classification. IEEE Trans. Geosci. Remote Sens. **58**(7), 5077–5091 (2020)
82. Q. Liu, L. Xiao, J. Yang, Z. Wei, CNN-enhanced graph convolutional network with pixel-and superpixel-level feature fusion for hyperspectral image classification. IEEE Trans. Geosci. Remote Sens. **59**(10), 8657–8671 (2020)
83. B. Liu, K. Gao, A. Yu, W. Guo, R. Wang, X. Zuo, Semisupervised graph convolutional network for hyperspectral image classification. J. Appl. Remote. Sens. **14**(2), 026516 (2020)
84. M. Lin, W. Jing, D. Di, G. Chen, H. Song, Context-aware attentional graph U-Net for hyperspectral image classification. IEEE Geosci. Remote Sens. Lett. **19**, 1–5 (2021)
85. F. Scarselli, M. Gori, A.C. Tsoi, M. Hagenbuchner, G. Monfardini, The graph neural network model. IEEE Trans. Neural Netw. **20**(1), 61–80 (2008)
86. M. Kampffmeyer, Y. Chen, X. Liang, H. Wang, Y. Zhang, E.P. Xing, *Rethinking Knowledge Graph Propagation for Zero-Shot Learning* (2018)
87. S. Zhang, S. Yan, X. He, *LatentGNN: Learning Efficient Non-local Relations for Visual Recognition* (2019)
88. X. Wang, H. Ji, C. Shi, B. Wang, P. Cui, P. Yu, Y. Ye, *Heterogeneous Graph Attention Network* (2019)
89. M.W.T.N. Kipf, *Semi-Supervised Classification with Graph Convolutional Networks* (2016)
90. Y. Li, D. Tarlow, M. Brockschmidt, R. Zemel, *Gated Graph Sequence Neural Networks* (2015)
91. L. Chi, G. Tian, Y. Mu, L. Xie, Q. Tian, Fast non-local neural networks with spectral residual learning, in *Proceeedings of the 27th ACM International Conference* (2019)

92. A. Vaswani, N. Shazeer, N. Parmar, J. Uszkoreit, L. Jones, A.N. Gomez, L. Kaiser, I. Polosukhin, *Attention Is All You Need* (2017)
93. W. Meng, W. Fu, S. Hao, H. Liu, X. Wu, Learning on big graph: label inference and regularization with anchor hierarchy. IEEE Trans. Knowl. Data Eng. **5**, 1 (2017)
94. A. Qin, Z. Shang, J. Tian, Y. Wang, T. Zhang, Y.Y. Tang, Spectral–spatial graph convolutional networks for semisupervised hyperspectral image classification. IEEE Geosci. Remote Sens. Lett. **16**(2), 241–245 (2018)
95. J. Chen, L. Jiao, X. Liu, L. Li, F. Liu, S. Yang, Automatic graph learning convolutional networks for hyperspectral image classification. IEEE Trans. Geosci. Remote Sens. **60**, 1–16 (2021)
96. A. Sha, B. Wang, X. Wu, L. Zhang, Semisupervised classification for hyperspectral images using graph attention networks. IEEE Geosci. Remote Sens. Lett. **18**(1), 157–161 (2020)
97. S. Wan, C. Gong, P. Zhong, B. Du, L. Zhang, J. Yang, Multiscale dynamic graph convolutional network for hyperspectral image classification. IEEE Trans. Geosci. Remote Sens. **58**(5), 3162–3177 (2019)
98. Y. Ding, X. Zhao, Z. Zhang, W. Cai, N. Yang, Multiscale graph sample and aggregate network with context-aware learning for hyperspectral image classification. IEEE J. Select. Top. Appl. Earth Observ. Remote Sens. **14**, 4561–4572 (2021)
99. Y. Cai, Z. Zhang, P. Ghamisi, Y. Ding, X. Liu, Z. Cai, R. Gloaguen, Superpixel contracted neighborhood contrastive subspace clustering network for hyperspectral images. IEEE Trans. Geosci. Remote Sens. (2022)
100. Y. Zhang, G. Cao, B. Wang, X. Li, P.Y.O. Amoako, A. Shafique, Dual sparse representation graph-based copropagation for semisupervised hyperspectral image classification. IEEE Trans. Geosci. Remote Sens. **60**, 1–17 (2021)
101. Y. Ding, Z. Zhang, X. Zhao, D. Hong, W. Cai, C. Yu, N. Yang, W. Cai, Multi-feature fusion: graph neural network and CNN combining for hyperspectral image classification. Neurocomputing (2022)
102. Y. Ding, Z. Zhang, X. Zhao, D. Hong, W. Li, W. Cai, Y. Zhan, AF2GNN: graph convolution with adaptive filters and aggregator fusion for hyperspectral image classification. Inform. Sci. **602**, 201–219 (2022)
103. M. Gori, G. Monfardini, F. Scarselli, A new model for learning in graph domains, in *Proceedings of the 2005 IEEE International Joint Conference on Neural Networks* (2005), pp. 729–734
104. C. Gallicchio, A. Micheli, Graph echo state networks, in *The 2010 International Joint Conference on Neural Networks (IJCNN)* (2010), pp. 1–8
105. J. Bruna, W. Zaremba, A. Szlam, Y. LeCun, *Spectral Networks and Locally Connected Networks on Graphs* (2013)
106. M. Henaff, J. Bruna, Y. Lecun, Deep convolutional networks on graph-structured data. Comput. Sci. (2015)
107. A. Micheli, Neural network for graphs: a contextual constructive approach. IEEE Trans. Neural Netw. **20**(3), 498–511 (2009)
108. J. Atwood, D. Towsley, Diffusion-convolutional neural networks. Comput. Sci. (2015)
109. M. Niepert, M. Ahmed, K. Kutzkov, Learning convolutional neural networks for graphs, in *International Conference on Machine Learning, PMLR* (2016), pp. 2014–2023
110. J. Gilmer, S.S. Schoenholz, P.F. Riley, O. Vinyals, G.E. Dahl, Neural message passing for quantum chemistry, in *International conference on machine learning. PMLR* (2017), pp. 1263–1272
111. Z. Wu, S. Pan, F. Chen, G. Long, C. Zhang, P.S. Yu, A comprehensive survey on graph neural networks. IEEE Trans. Neural Netw. Learn. Syst. **1**, 32 (2021)
112. D.I. Shuman, S.K. Narang, P. Frossard, A. Ortega, P. Vandergheynst, The emerging field of signal processing on graphs: extending high-dimensional data analysis to networks and other irregular domains. IEEE Sig. Process. Mag. **30**(3), 83–98 (2013)
113. S.G. Mallat, *A Wavelet Tour of Signal Processing: The Sparse Way* (China Machine Press, 2009)

114. L. Li, Z. Gan, Y. Cheng, J. Liu, Relation-aware graph attention network for visual question answering, in *Proceedings of the IEEE/CVF International Conference on Computer Vision* (2019), pp. 10313–10322
115. J. Gilmer, S.S. Schoenholz, P.F. Riley, O. Vinyals, G.E. Dahl, Neural message passing for quantum chemistry, in *International Conference on Machine Learning, PMLR* (2017)
116. G. Li, M. Müller, G. Qian, I.C. Delgadillo, A. Abualshour, A. Thabet, B. Ghanem, DeepGCNs: making GCNs go as deep as CNNs. IEEE Trans. Pattern Anal. Mach. Intell. (2019)
117. Y. Wang, Y. Sun, Z. Liu, S.E. Sarma, M.M. Bronstein, J.M. Solomon, Dynamic graph cnn for learning on point clouds. ACM Trans. Graph. **38**(5), 1–12 (2019)
118. P. Veličković, G. Cucurull, A. Casanova, A. Romero, P. Lio, Y. Bengio, *Graph Attention Networks* (2017)
119. N. Peng, H. Poon, C. Quirk, K. Toutanova, W.-T. Yih, Cross-sentence n-ary relation extraction with graph LSTMS. Trans. Assoc. Comput. Lingu. **5**, 101–115 (2017)
120. D.K. Duvenaud, D. Maclaurin, J. Iparraguirre, R. Bombarell, T. Hirzel, A. Aspuru-Guzik, R.P. Adams, Convolutional networks on graphs for learning molecular fingerprints. Adv. Neural Inform. Process. Syst. **28**, 598 (2015)

Chapter 2
Graph Sample and Aggregate-Attention Network for Hyperspectral Image Classification

2.1 Introduction

Hyperspectral images (HSIs) provide detailed spectral information through hundreds of (narrow) spectral channels, which can be used to accurately classify diverse materials of interest [1, 2]. However, the increased dimensionality of such data provides a challenge to conventional techniques, and hyperspectral classification has great research value.

In the past few decades, people have conducted significant efforts on HSI classification, which can be summarized into two categories: traditional methods and neural network methods. Traditional methods have made some efforts on explored more discriminative feature representations, such as morphological features and texture features [3]. Apart from these subspaces learning, sparse learning algorithms and machine learning, such as random forest and support vector machine (SVM) [4, 5] have got great attention in the community. However, traditional methods have defects in feature extraction completeness and may suffer from overfitting because of the deficiency in training samples.

Deep learning has achieved great success in many applications and it has also greatly promoted the technological progress of HSI classification. For example, in [6, 7], the HSIs were classified using different dimensional convolutions; In Liu et al. [8], proposed introduced a spectral-spatial featural extraction method based on LSTM network. In Zhang et al. [9], used image semantic context to classify HSIs. Ref. [10] adopted residual networks to learn spatial and spectral charact-eristics of the image to improve the classification rate. In [11], an unsupervised spectral-spatial featural extraction network was proposed. However, CNN needs a large number of training labels and calculation. Simultaneously, the CNN only performs convolution on the regular region. Furthermore, the size of the CNN convolution kernel is fixed, which will lead to edges missing phenomenon in the process of feature extraction [12].

Y. Ding et al., *Graph Neural Network for Feature Extraction and Classification of Hyperspectral Remote Sensing Images*, Intelligent Perception and Information Processing, https://doi.org/10.1007/978-981-97-8009-9_2

To ameliorate these issues, people have conducted extensive researches on classification using GCNs. The GCN conducts semisupervised learning on graph-structured data and can operate on graph signals directly via a variant of CNNs. Sha et al. [13] applied the graph attention network to hyperspectral classification. Mou et al. [14] proposed a nonlocal graph convolution network, which constructs a graph by calculating the relationship between non-adjacent pixels to improve the classification accuracy. Hong et al. [15] proposed a graph convolution classification method combining GCN and CNN to increase the classification accuracy. In Wan et al. [16], employed a multiscale graph convolutional network to extract multiscale graph features. Ref. [17] adopted a context-aware mechanism to learn the local contextual of the graph. The mentioned methods are GCN-based. However, GCN is a transductive learning and whole graph training method, which is difficult to aggregate the new node and will bring a huge amount of computation.

The main contributions in this letter are as follows: (1) the incorporation of SAGE (first time) for extracting contextual relations among superpixels; (2) the utilization of multi-level graph projection and flexible reprojection framework for extracting long-range contextual relations and producing truthful local-region features; (3) the adoption of attention mechanism graph refinement for characterizing global and contextual relations and accurately finding precise region representations.

2.2 Related Work

Many researchers have published their methods to classify HSIs. In this part, we mainly introduce the graph neural network (GNN) method, which has a lot of relationships with our work.

2.2.1 Spectral Graph Convolution

For a graph $\boldsymbol{G} = (\boldsymbol{V}, \boldsymbol{E})$, the Laplacian matrix $\boldsymbol{L}$ is expressed as

$$\boldsymbol{L} = \boldsymbol{D} - \boldsymbol{A} \tag{2.1}$$

where $\boldsymbol{V}$ represents the vertex set and $\boldsymbol{E}$ represents the edge set, $\boldsymbol{D}$ is the degree matrix of the graph, and $\boldsymbol{A}$ is the adjacency matrix of the graph.

The corresponding symmetrically normalized Laplacian $\boldsymbol{L}_n$

$$\boldsymbol{L}_n = \boldsymbol{I} - \boldsymbol{D}^{\frac{1}{2}}\boldsymbol{A}\boldsymbol{D}^{-\frac{1}{2}} \tag{2.2}$$

where $\boldsymbol{I}$ is an identity matrix.

To conduct node embedding for $\boldsymbol{G}$, a filter $\boldsymbol{g}_{\theta} = \text{diag}(\boldsymbol{\theta})$ parameterized by $\boldsymbol{\theta} \in \boldsymbol{R}^N$ is defined, which can be expressed as the multiplication of a signal $\boldsymbol{x}$ (a scalar for each node) with $\boldsymbol{g}_{\theta}$ the Fourier domain, that is

$$\boldsymbol{g}_{\theta} * \boldsymbol{x} = \boldsymbol{U}\boldsymbol{g}_{\theta}(\boldsymbol{\Lambda})\boldsymbol{U}^T\boldsymbol{x} \tag{2.3}$$

where $\boldsymbol{U}$ is the matrix of eigenvectors of $\boldsymbol{L}_n$ and can be computed by $\boldsymbol{L}_n = \boldsymbol{U}\Lambda\boldsymbol{U}^{-1}$. $\boldsymbol{\Lambda}$ is the diagonal matrix of eigen-values of $\boldsymbol{L}_n$. $\boldsymbol{g}_{\theta}$ can be understood as a function of eigen-values of $\boldsymbol{\Lambda}$, i.e., $\boldsymbol{g}_{\theta}(\boldsymbol{\Lambda})$. . However, note that evaluating (3) requires explicitly calculating. Hammond et al. approximately fitted $\boldsymbol{g}_{\theta}(\boldsymbol{\Lambda})$ by using a truncated expansion in terms of Chebyshev polynomials up to K th order, which leads to

$$\boldsymbol{g}_{\theta}(\boldsymbol{\Lambda}) \approx \sum_{k=0}^{K} \boldsymbol{\theta}_k \boldsymbol{T}_k(\boldsymbol{L}_n) \tag{2.4}$$

where $\boldsymbol{T}_k$ is the Chebyshev polynomials.

Kipf and Welling simplified Eq. (2.4) K = 1, which is the first order approximation by Chebyshev polynomials, and largest eigenvalue $\lambda_{max} \approx 2$. Equation (2.4) can be rewritten as

$$\boldsymbol{g}_{\theta} * \boldsymbol{x} \approx \boldsymbol{\theta}(\boldsymbol{I}_N + \boldsymbol{L})\boldsymbol{x} = \boldsymbol{\theta}\left(\tilde{\boldsymbol{D}}^{\frac{1}{2}} \tilde{\boldsymbol{A}} \tilde{\boldsymbol{D}}^{-\frac{1}{2}}\right)\boldsymbol{x} \tag{2.5}$$

With the activation layer σ, the GCN propagation rules as follows

$$\boldsymbol{H}^{(l+1)} = \sigma\left(\tilde{\boldsymbol{D}}^{\frac{1}{2}} \tilde{\boldsymbol{A}} \tilde{\boldsymbol{D}}^{-\frac{1}{2}} \boldsymbol{H}^{(l)}\boldsymbol{W}^{(l)}\right) \tag{2.6}$$

where $\boldsymbol{H}^{(l+1)}$, $\boldsymbol{H}^{(l)}$ denote the values of $l + 1$ and l layers respectively, and $\boldsymbol{W}$ is the weight matrix.

2.2.2 Spatial Graph Convolution

According to the traditional CNN operation on the image, the spatial based GNN defines the graph convolution operator based on the spatial relationship of a node. The image is regarded as a special graph, and each pixel represents a node, because the adjacent nodes have a fixed order, the training weights can be shared with different local spaces. Spatial based GNNs method has better efficiency, flexibility, and versatility compared with GCN. If you are interested in the details of these algorithms, the readers can refer to relevant papers.

2.3 Proposed Method

In this section, we will present SAGE-A for HSI classify-cation (Fig. 2.1). It is mainly composed of three parts, including pixel-to-Region assignment (3.1), contextual relations refinement (3.2, 3.3), and region-to-pixel assignment (3.4).

2.3.1 Pixel-to-Region Assignment

HSI contains a large number of pixels in the spatial dimension, and a huge amount of computation is needed to classification and sometimes it is unacceptable. To ameliorate this issue, we find that neighbor pixels have a large probability to belong to the same land-cover type. Therefore, SLIC has been adopted to segment the entire image into a small number of local regions, and the pixels consisted of regions that have a strong spectral-spatial similarity. Concretely, SLIC conducts image region segmentation via iteratively growing the local clusters by employing a k-means algorithm. In the letter, the local regions are treated as the graph node, which can significantly reduce the number of graph nodes and improve computational efficiency. Here, the average spectral signatures of the involved pixels in the node (local region) are taken as the feature vector of the node.

2.3.2 GraphSAGE

Traditional GCN models assume that the adjacent pixels on a graph are more likely to share the same label, thus the label information can be propagated to unlabeled samples from labeled samples via graph Laplacian regularization. However, transductive learning requires all nodes to participate in training to get node embedding,

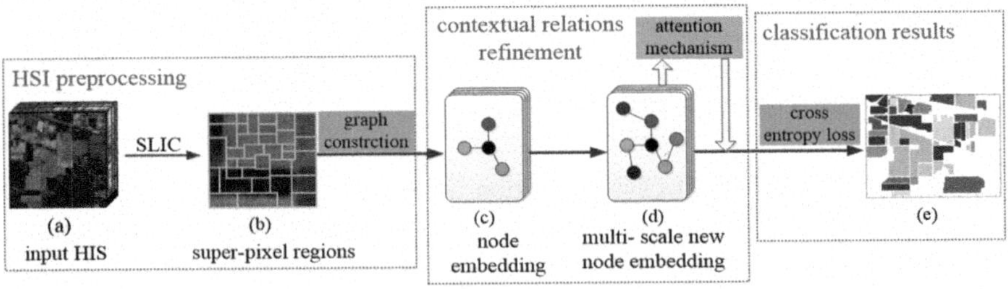

Fig. 2.1 An overview of the SAGE-A network. **a** Is the HSI. **b** The superpixels segmented by the SLIC algorithm. In **c**, the circles and lines represent the superpixel (graph node) and edges. Different colors of the nodes represent different land-cover types and the input of the network are spectral characteristics of each node. **d** Denotes multi-level new node embedding using SAGE and attention mechanism

and it cannot quickly get embedding of new nodes. Therefore, GCN can only learn the information about neighboring nodes, and cannot naturally generalize to the unknown vertices. In this letter, a graphSAGE algorithm is adopted to learn more spatial scale information, which improves the generalization ability of the model for new nodes. The forward propagation rule is expressed as Algorithm 1.

Algorithm 1: GraphSAGE embedding generation (i.e., forward propagation) algorithm.

Input: Graph $\boldsymbol{G} = (\boldsymbol{V}, \boldsymbol{E})$; input features $\{\boldsymbol{x}_v, \forall v \in \boldsymbol{V}\}$; the number of layers of the network K; weight matrices $\boldsymbol{W}^k, \forall k \in \{1, \cdots, K\}$; non-linearity σ; mean aggregator functions AGG; neighborhood function $N : v \to 2^v$

Output: Vector representations for all $v \in \boldsymbol{V}$

1: $h_0 \leftarrow \boldsymbol{x}_v, \forall v \in \boldsymbol{V}$;
2: for $k = \{1, \cdots, K\} do$
3: for $v \in V$ do
4: $\boldsymbol{h}^k_{N(v)} \leftarrow AGG\left(\left\{\boldsymbol{h}^{k-1}_u, \forall u \in \boldsymbol{N}(v)\right\}\right)$;
5: $\boldsymbol{h}^k_v \leftarrow \sigma\left(\boldsymbol{W}_k \cdot CONCAT\left(\boldsymbol{h}^{k-1}_v, \boldsymbol{h}^k_{N(v)}\right)\right)$
6: end
7: $\boldsymbol{h}^k_v \leftarrow \frac{\boldsymbol{h}^k_v}{\|\boldsymbol{h}^k_v\|_2}, v \in V$
8: end

Output: $z_v \leftarrow \boldsymbol{h}^K_v, v \in V$

In Algorithm 1, K is the number of layers of the network, which also represents the number of hops of adjacent points that can be aggregated at each vertex, because each additional layer can aggregate the information about the neighbors of a further layer, $\forall u$ is the eigenvector of the node u, $\left\{\boldsymbol{h}^{k-1}_u, \forall u \in \boldsymbol{N}(v)\right\}$ denotes the embedding of the neighbor U of the node V in the $k - 1$ layer,$\boldsymbol{h}^k_v$ represents the characteristic of all neighbors of node v at the k level. In this letter, AGG can be expressed as $AGG = \sum_{u \in N(v)} \frac{\boldsymbol{h}^{k-1}_u}{|N(v)|}$. From Algorithm 1, each iteration or search depth, nodes collect information from their local neighbors, and as the process iterates, nodes gradually get more and more information from farther reaches of the graph. Thus, long-range contextual relations are extracted.

2.3.3 Graph Attention

In the experiment, we find that the association degree between different nodes is different. To extract the global and contextual information better, a graph attention mechanism is added into the network to make the important node information have greater weight. The graph attention mechanism can obtain the global geometric features by calculating the relationship between any two nodes in the graph. To get the corresponding transformation between input and output, it is necessary to obtain the output features by linear transformation according to the input characteristics at least once. A weight matrix is trained for all nodes: $W \in R^{F\prime \times F}$, which is the relationship between the input features F and the output features $F\prime$. Node to node correlation can be learned through the network layer

$$e_{ij} = \left(LeakyReLU\left(\boldsymbol{a}^T\left[W\boldsymbol{h}_i||W\boldsymbol{h}_j\right]\right)\right) \tag{2.7}$$

Equation (2.7) shows the importance of node j to node$i$, $a^T \in R^{2F}$ is the parameter vector of the network, || denotes concatenation operation, and $LeakyReLU(\cdot)$ is a nonlinear layer.

Then normalizing and converting e_{ij} to a probability output a_{ij} through a softmax function

$$a_{ij} = \frac{\exp\left(LeakyReLU\left(\boldsymbol{a}^T\left[\boldsymbol{Wh}_i||\boldsymbol{Wh}_j\right]\right)\right)}{\sum_{k\in N_i}\exp\left(LeakyReLU\left(\boldsymbol{a}^T\left[\boldsymbol{Wh}_i||\boldsymbol{Wh}_j\right]\right)\right)} \tag{2.8}$$

Therefore, the graph convolution output of each node can be expressed as follows

$$\boldsymbol{h}_i^l = \sigma\left(\sum_{j\in N_i} a_{ij}\cdot \mathbf{W}^T\boldsymbol{h}_i^{l-1}\right) \tag{2.9}$$

where σ denotes the activate function,a_{ij} is learned attention weight.

Figure 2.2 shows the working process of the graph attention mechanism in SAGE-A. By learning the importance weight of each node to the classified node, the graph attention mechanism makes the important nodes have greater weight, hence global and contextual information can be learned from the graph via an attention mechanism.

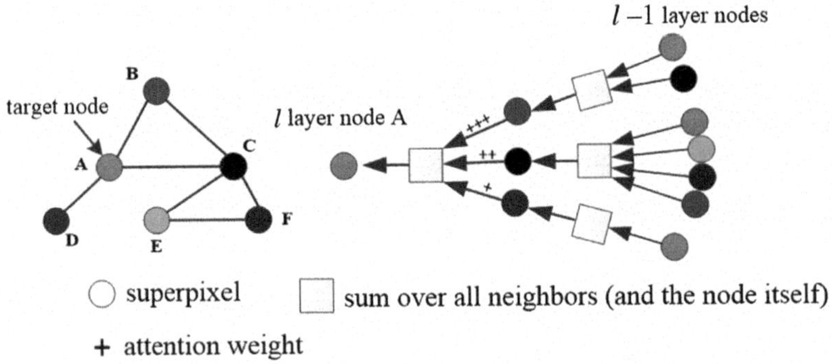

Fig. 2.2 Schematic diagram of graph attention mechanism (Eq. (2.9)). l denotes the lth layer of the network. Different colors of the nodes represent different land-cover types feature

2.3.4 Region-to-Pixel Assignment

Multiscale information has been widely proved to be very useful for HSI classification [18, 19]. Ground objects have different geometric features, and multi-level feature extraction can fully learn the contextual information of the image. The network uses multi-layer graphSAGE to learn the relationship between superpixels of different scales as Algorithm 1. Figure 2.3 demonstrates 1-hop and 2-hop neighbors of a central example A to illustrate the multis-level design as. Then, the receptive field of A at the scale s is formed as

$$\boldsymbol{H}_s(x_i) = \boldsymbol{H}_1(\boldsymbol{H}_{s-1}(\boldsymbol{x}_i), \boldsymbol{x}_{s-1}) \tag{2.10}$$

Where $\boldsymbol{H}_0(\boldsymbol{x}_i) = \boldsymbol{x}_i$,$\boldsymbol{H}_1(\boldsymbol{x}_i)$ is the new node embedding of 1-hop neighbors of $\boldsymbol{x}_i$. Considering that the information association degree of different nodes is different, we use Eq. (2.10) to analyze the association degree of the learned information. The network output is expressed as follows

$$\boldsymbol{O} = \boldsymbol{A}(\boldsymbol{H}_s(\boldsymbol{x}_i)) \tag{2.11}$$

where $\boldsymbol{A}$ is the attention mechanism, and $\boldsymbol{O}$ is the output of SAGE-A. In our network, the cross-entropy error is adopted to penalize the difference between the network output and the labels of the original labeled examples, namely

$$\boldsymbol{L} = -\sum_{z \in \mathbf{y}_G} \sum_{f=1}^{C} \boldsymbol{Y}_{zf} \, ln \boldsymbol{O}_{zf} \tag{2.12}$$

where $\mathbf{y}_G$ is labeled examples set, C denotes the number of classes and $\boldsymbol{Y}_{zf}$ is the label matrix. The details of our SAGE-A are shown in Algorithm 2. The input feature of the SAGE-A is the average spectral signatures of the graph nodes, which enables the network to process the spectral information about HSIs. At the same time, the SAGE method is adopted to process the spatial relationship of the nodes in the graph network, so that the model can learn the long-range spatial information of the HSIs, the graph attention mechanism is used to process the overall information of the graph to learn the global and contextual information of each node in the graph.

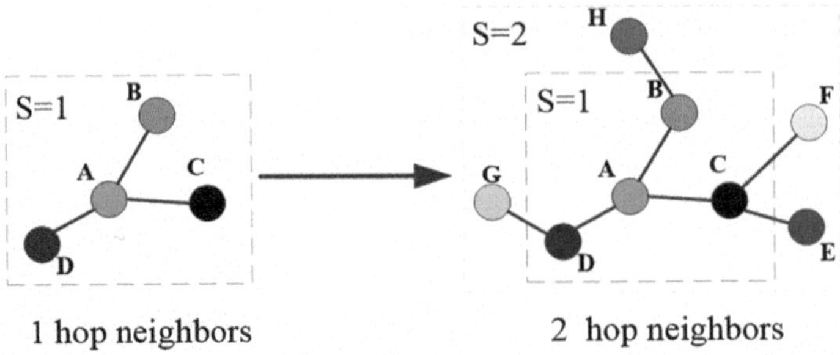

Fig. 2.3 Network multilevel information learning

2.4 Experimental Results

2.4.1 Dataset Description and Implementation

Two real hyperspectral datasets of Pavia University (PU) and Houston 2013 are adopted to verify the classification performance of our proposed method. The first dataset PU contains 610 × 340 pixels and 103 bands, including a large number of background pixels, and 42,776 pixels can be applied to classification. The whole map contains 9 kinds of features. The second dataset Houston 2013 has been used in the 2013 GRSS Data Fusion Contest. The Salinas scene is composed of 144 spectral bands and 349 × 1905 pixels. These pixels are divided into 15 categories.

Algorithm 2: The Proposed SAGE-A for HSI Classification

Input: Input image; number of epoch T=2000; learning rate = 0.0001; number of graph convolutional layers L=3; dropout=0.2; Adam gradient descent; python=3.8; pytorch=1.6.0.

Segment the whole image into super-pixels via SLIC algorithm;
Extract the superpixels input features (average spectral signatures);
//Construct the graph and train the SAGE-A model
for t = 1 to T **do**
  Graph convolution notes feature;
  Perform graph learning at adjacent points spatial feature by Algorithm 1;
  Bachnormalzation, dropout and relu;
  Perform graph learning at adjacent points and farther points spatial level by Algorithm 1;
  Bachnormalzation and relu;
  //Graph convolution attention mechanism
  Perform graph learning at global level by Eq. (2.9) and output the spatial and spectral feature;
  Calculate the error term according to Eq. (2.12) and update the weight matrices using Adam gradient descent;
end for
Conduct label prediction based on the trained network;

Output: Predicted label for each pixel.

2.4.2 Experimental Setting

For the two HSI datasets described in Sect. 2.4.1, 30 labeled pixels in each class are randomly selected for network training, and the remaining unlabeled pixels are used for network testing. The hyperparameters selection in our SAGE-A is shown in Algorithm 2. To validate the performance of SAGE-A, the other six state-of-the-art image classification methods are employed to conduct a comparison. Specifically, our network is compared with two CNN-based methods, i.e., convolution autoencoder (CAE) [11] and convolutional RNN (CRNN) [8], and two GCN-based methods, i.e., context-aware dynamic graph convolutional network (CAD-GCN) [17] and multi-scale dynamic graph convolutional network (MDGCN) [16]. Meanwhile, two traditional machine learning methods are also adopted, namely multiband compact texture unit (MBCTU) [3] and Joint collaborative representation and SVM with Decision Fusion (JSDF) [4]. Overall accuracy (OA), average accuracy (AA), kappa coefficient (κ), and per-class accuracy are adopted to are used as evaluation indices.

2.4.3 Comparisons with Other Methods

From Tables 2.1 and 2.2, we conclude that the proposed SAGE-A achieves better results compared with the other state-of-the-art models in OA, AA and κ, which validates the effectiveness of the proposed multi-level graphSAGE network with an attention mechanism. It is also notable that the GCN-based methods perform better than MBCUT, JSDF, CAE, CRNN. This is because the graph convolution network can learn the relations among neighbor nodes automatically, which is suit for classification with limited labeled training samples. However, the GCN is a transductive learning method, which is difficult to aggregate the new node. In another word, the GCN-based methods are difficult to learn the long-range contextual information of the graph. While our proposed SAGE-A aggregate the different levels of nodes (superpixel) via adjusting the node embedding layers of the network. And, the employment of an attention mechanism enables the SAGE-A to automatically learn the global and contextual information of the graph.

2.4.4 Impact of Parameters/Hyperparameters

Many significant parameters/hyperparameters should be tuned in the proposed SAGE-A architecture. In the experiment, the sensitivity of the classification performance to different hyperparameter settings will be evaluated in detail.

Figure 2.4 demonstrates the classification performances of the seven algorithms with different numbers of labeled examples (i.e., pixels) for training are investigated. we vary the number of labeled examples per class from 5 to 30 with an interval of

Table 2.1 Accuracy comparisons for the Pavia university scene

ID	MBCUT	JSDF	CAE	CRNN	CAD-GCN	MDGCN	SAGE-A
1	87.49	82.40	84.78	79.85	83.85	93.55	**96.50**
2	89.11	90.76	89.34	82.33	95.56	**99.25**	99.17
3	86.24	86.71	82.96	89.67	95.50	92.03	**98.26**
4	90.61	92.88	86.32	91.45	81.90	83.78	92.85
5	97.18	**100.00**	97.65	94.12	98.91	99.47	97.49
6	93.25	94.30	91.33	91.37	**97.81**	95.26	94.78
7	93.49	96.62	94.49	93.67	96.79	98.92	**97.92**
8	84.14	94.69	81.32	80.18	**95.35**	94.99	89.84
9	96.57	**99.56**	80.11	82.34	84.80	81.03	98.80
OA	89.43	90.82	86.81	85.46	92.91	95.68	**96.19**
AA	90.90	93.10	87.59	87.22	92.27	93.15	**96.18**
Kappa	0.86	0.88	0.86	0.84	0.90	0.94	**0.96**

Bold numbers indicate the best performance

Table 2.2 Accuracy comparisons for the houston 2013 scene

ID	MBCUT	JSDF	CAE	CRNN	CAD-GCN	MDGCN	SAGE-A
1	92.86	**97.41**	81.54	82.45	94.45	93.42	86.17
2	92.18	**99.84**	81.67	84.12	96.43	93.67	94.42
3	97.42	**99.88**	77.31	91.56	95.17	98.12	98.80
4	90.96	**98.22**	91.53	91.29	94.82	95.58	96.19
5	97.17	**100**	95.85	98.81	98.91	99.00	**100.00**
6	91.78	**99.32**	97.76	94.83	97.48	93.28	93.72
7	82.88	91.32	86.94	86.42	91.58	87.68	**93.31**
8	71.85	68.82	67.62	53.05	74.63	80.45	**84.63**
9	81.94	69.47	85.4	84.04	86.75	89.64	**96.74**
10	87.31	85.63	57.85	45.14	94.23	90.06	**97.59**
11	77.41	94.51	72.92	61.85	**94.65**	86.73	89.39
12	86.35	84.33	**94.02**	84.40	89.55	89.44	92.98
13	85.58	98.10	89.68	84.14	**96.80**	92.78	85.69
14	96.85	**100.00**	97.73	96.03	**100.00**	99.43	**100.00**
15	92.27	**99.86**	74.71	93.45	98.02	96.27	98.46
OA	87.07	90.51	83.50	82.10	92.51	91.40	**92.76**
AA	88.32	92.46	81.84	79.21	93.57	92.37	**93.87**
Kappa	0.8601	0.8974	0.80	0.78	0.92	0.91	**0.93**

Bold numbers indicate the best performance

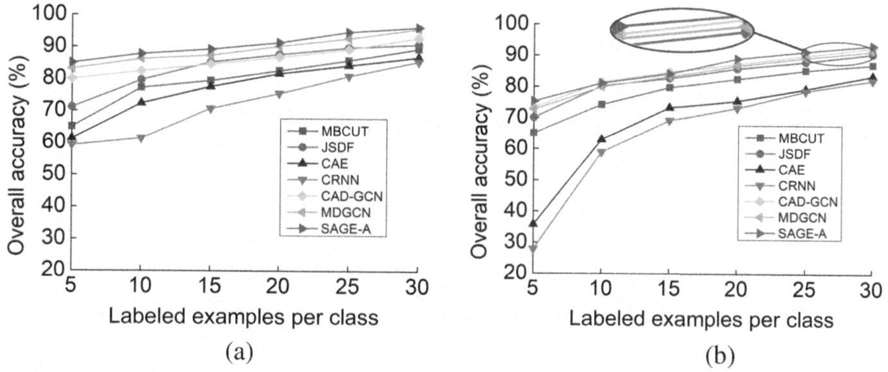

Fig. 2.4 Overall accuracies of various methods under different numbers of labeled examples per class. **a** University of Pavia dataset. **B** Houston 2013 dataset

5 and report the OA performance acquired by seven algorithms on PU, Houston 2013 datasets. From the results, we can find that the OA of each proposed method in PU and Houston 2013 datasets has been significantly improved with the increase of labeled examples per class. Besides, the proposed SAGE-A model performs better than the contrast algorithms from beginning to end, which shows the effectiveness of multi-level spatial information on HSI classification. Furthermore, the proposed SAGE-A allows to automaticity learn global contextual features based on classified land-cover, which is more robust than using a pre-computed fixed graph.

The impact of convolution layers l and segment scales S on the two data sets is revealed in Fig. 2.5. We can conclude that both l and S have a significant impact on the classification accuracies. Meanwhile, the best result is usually reached with convolution layer 3. Multi-level is able to learn more spatial information at a larger scale. However, the characteristics learned through iteration or search depth have an inhibitory effect on the classification due to the low correlation. For the S, the classification accuracies would increase as the segment scale increasing. However, the amount of calculation also increases exponentially, which may be unaccepted under limited experimental conditions. In our proposed method, the segment scales S is 30,000, which has reached the limits of computing.

2.4.5 Ablation Study

In this experiment, we investigate the ablative effect of the SAGE-based attention mechanism. For the sake of comparison, we record the classification results produced without using an attention mechanism, and the simplified model is denoted

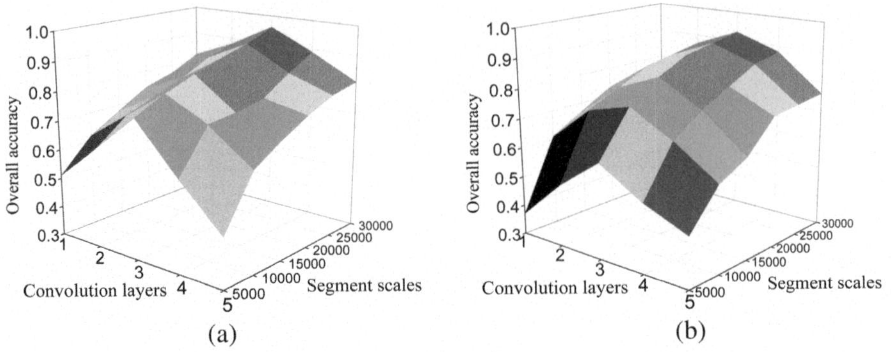

Fig. 2.5 Parametric sensitivity of l and S. **a** Pavia University dataset. **b** Houston 2013 dataset

Table 2.3 OA, AA (%), and Kappa coefficient achieved by different model settings on PU and houston 2013 datasets

Dataset	Pavia University		Houston 2013	
Methods	SAGE	SAGE-A	SAGE	SAGE-A
OA	86.64	96.19	83.41	92.76
AA	86.79	96.18	84.04	93.87
Kappa	0.87	0.96	0.84	0.93

as ‘SAGE’. And the experiment setting is kept identical to Sect. 2.4.2. The comparative results are demonstrated in Table 2.3. As shown in the table, the SAGE-based attention mechanism plays an important role in the improvement of learning efficiency.

2.5 Conclusion

In this letter, a novel SAGE-A for HSI classification is proposed. To extract long-range contextual relations, we go beyond the regular image grids by adopting the pixel-to region (superpixel) assignment, and further encode the contextual relations among local-regions, so that originally long-range local-node in the 2-D space can be connected by multi-level SAGE. Moreover, we learn the importance weight of each node to the classified node, Therefore, global and contextual relations among pixels can be gradually refined, and local-region features can be represented precisely.

References

1. B. Rasti et al., Feature extraction for hyperspectral imagery: the evolution from shallow to deep—overview and toolbox. IEEE Geosci. Remote Sens. **8**(4), 60–88 (2020)
2. P. Zhong, Z. Gong, J. Shan, Multiple instance learning for multiple diverse hyperspectral target characterizations. IEEE Trans. Neural Netw. Learn. Syst. **12**, 1–13 (2019)
3. K. Djerriri, A. Safia, R. Adjoudj, M.S. Karoui, Improving hyperspectral image classification by combining spectral and multiband compact texture features, in *Proceedings of the IEEE International Geosciences Remote Sensing Symposium (IGARSS)* (2019), pp. 465–468
4. C. Bo, H. Lu, D. Wang, Hyperspectral image classification via JCR and SVM models with decision fusion. IEEE Geosci. Remote Sens. Lett. **13**(2), 177–181 (2016)
5. L. Wang, S. Hao, Q. Wang, Y. Wang, Semi-supervised classification for hyperspectral imagery based on spatial-spectral label propagation. ISPRS J. Photogram. Remote Sens. **97**, 123–137 (2014)
6. W. Hu, Y. Huang, L. Wei, F. Zhang, H. Li, Deep convolutional neural networks for hyperspectral image classification. J. Sens. **1–12**, 2015 (2015)
7. K. Makantasis, K. Karantzalos, A. Doulamis, N. Doulamis, Deep supervised learning for hyperspectral data classification through convolutional neural networks, in *Proceedings of the IEEE International Geosciences Remote Sensing Symposium (IGARSS)* (2015), pp. 4959–4962
8. Q. Liu, F. Zhou, R. Hang, X. Yuan, Bidirectional-convolutional lstm based spectral-spatial feature learning for hyperspectral image classification. Remote Sens. **9**(12), 1330 (2017)
9. M. Zhang, W. Li, Q. Du, Diverse region-based CNN for hyperspectral image classification. IEEE Trans. Image Process. **27**(6), 2623–2634 (2018)
10. K. He, X. Zhang, S. Ren, J. Sun, Deep residual learning for image recognition, in *Proceedings of the IEEE Conference on Computer Vision Pattern Recognition (CVPR)* (2016), pp. 1–9
11. R. Kemker, C. Kanan, Self-taught feature learning for hyperspectral image classification. IEEE Trans. Geosci. Remote Sens. **55**(5), 2693–2705 (2017). https://doi.org/10.1109/TGRS.2017.2651639
12. D. Hong, N. Yokoya, J. Chanussot, X.X. Zhu, An augmented linear mixing model to address spectral variability for hyperspectral unmixing. IEEE Trans. Image Process. (2019)
13. A. Sha, B. Wang, X. Wu, L. Zhang, Semisu-pervised classification for hyperspectral images using graph attention networks. IEEE Geosci. Remote Sens. Lett. **8**, 23 (2020)
14. L. Mou, X. Lu, X. Li, X.X. Zhu, Nonlocal graph convolutional networks for hyperspectral image classification. IEEE Trans. Geosci. Remote Sens. **55**(5), 3626–3635 (2019). https://doi.org/10.1109/TGRS.2020.3015157
15. D. Hong, L. Gao, J. Yao, B. Zhang, A. Plaza, J. Chanussot, Graph convolutional networks for hyperspectral image classification. IEEE Trans. Geosci. Remote Sens. **55**, 5966–5978 (2020)
16. S. Wan, C. Gong, P. Zhong, B. Du, L. Zhang, J. Yang, Multiscale dynamic graph convolutional network for hyperspectral image classification. IEEE Trans. Geosci. Remote Sens. **58**(5), 3162–3177 (2020)
17. S. Wan, C. Gong, P. Zhong, S. Pan, G. Li, J. Yang, Hyperspectral image classification with context-aware dynamic graph convolutional network. IEEE Trans. Geosci. Remote Sens. **59**, 597–612 (2021)
18. N. Srivastava, G. Hinton, A. Krizhevsky, I. Sutskever, R. Salakhut-dinov, Dropout: a simple way to prevent neural networks from over fitting. J. Mach. Learn. Res. **15**(1), 1929–1958 (2014)
19. S. Zhang, S. Li, Spectral-spatial classification of hyperspectral images via multiscale super pixels based sparse representation, in *Proceedings of the IEEE IGARSS* (2016), pp. 2423–2426

Chapter 3
Multi-feature Fusion: Graph Neural Network and CNN Combining for Hyperspectral Image Classification

3.1 Introduction

Hyperspectral imagery collected from satellite or airborne comprises hundreds of contiguous bands and contains abundant spectral-spatial information. Due to the advantages of HSI, land-cover categories can be distinguished at the pixel level. Then HSI classification (classify each pixel into a certain label) has attracted extensive attention [1–3]. Hence, HSI classification has been applied to various fields, such as land management [4, 5], environmental monitoring [6], a military investigation [7, 8], and agriculture assessment [9, 10]. However, high dimensionality [11], labeled training samples deficiency [12] and complex spectral noise effects [13] bring great difficulties to HSI classification, and various methods have been researched to solve the aforementioned problems.

In the past several decades, various machine learning classification methods have been adopted for HSI classification [14]. In the early stage, the spectral and spatial information usually have been processed separately, and most of the classifiers aimed to separate the spectra in a high-dimensional space. For example, the extreme learning machine (ELM) [15], logistic regression [16], and support vector machine (SVM) [17] have been developed to classify the HSI pixel-wise. However, these methods usually focus on spectral dimension while ignoring the perspective of spatial information, and considerable errors or outliers have been produced in classified maps [18]. To extract the spatial information from HSI, various spatial-spectral-based methods considering both spectral and spatial contextual information have been proposed, e.g., morphological profiles (MPs) [19] and extended maldistributed profiles (EMAP) [20]. Furthermore, the Gabor filter [21, 22], wavelets [23], extended morphological and profiles filter [24], and edge-preserving filter [25] have been applied to HSI classification to extract the texture feature of HSI. The traditional machine learning methods have made great contributions to HSI classification. However, the aforementioned methods are quite empirical and heavily rely on professional expertise.

Y. Ding et al., *Graph Neural Network for Feature Extraction and Classification of Hyperspectral Remote Sensing Images*, Intelligent Perception and Information Processing, https://doi.org/10.1007/978-981-97-8009-9_3

In addition, the robustness and classification accuracy rate of traditional machine learning methods are inferior to deep learning methods.

In contrast with the traditional feature extraction techniques, deep learning methods can extract robust and adaptive deep features from labeled data automatically. Due to the powerful feature learning ability of deep learning, it has been widely adopted by extensive filed. Naturally, deep learning has also been well applied in the HSI classification task and has demonstrated excellent performance [6, 26]. For instance, recurrent neural networks (RNNs) [27], stacked auto-encoders (SAEs) [28], and the convolutional neural networks (CNNs) [29, 30] were proposed to extract deep features of HSI. Among these methods, CNN-based methods have become the most widely adopted framework for HSI classification. Furthermore, the different structures appeared one after another to extract deep spectral-spatial features from HSI, ranging from the 1D-CNN to the 3D-CNN, and from the single CNN to the hybrid CNN. For example, in [31], a 1-D CNN model was designed for extracting the spectral information of each pixel. However, the 1-D CNN cannot learn the spatial information from HSI, which limits its performance. Then, in [32], a 3D-CNN was proposed for learning spectral-spatial features. While, Yang et al. [33] proposed a two-channel CNN (TCCNN), integrating 1D CNN and 2D CNN. Furthermore, Chen et al. [34] improved TCCNN, added a 3D CNN branch, and a multichannel CNN (MCCNN) were designed. Now the latest CNN development results have also been applied to HSI classification, e.g., the residual and dense connections. In [35], the Fully Convolutional Network (FCN) was unutilized for deep features of HSI. Generative Adversarial Networks (GANs) [36] have also been widely investigated in HSI classification. However, the deep CNN methods are facing some limitations. Firstly, the recent deep CNNs are computational complexity and require a superior computing capacity. Furthermore, due to the requirement of the labeled samples, most CNNs are not suited for classification with limited labeled samples. Most importantly, the aforementioned deep learning methods are designed for Euclidean data, and the correlation between land types is ignored [37].

To overcome the drawbacks of CNNs for HSI classification, graph convolutional networks (GCNs) have attracted increasing interests. GCN [38] is a semi-supervised framework and can perform convolution operations on non-Euclidean data. The combination of HSI and GCN is one of the hotspots of current research. Qin et al. [39] firstly introduced a semi-supervised GCN method for HSI Classification, in which each pixel in the HSI was treated as a node, and then the GCN was adopted to process the graph. Bai et al. [40] developed a deep attention graph convolutional network (DAGCN) to extract the physical and chemical non-Euclidean structures of HSI for classification. However, the applications of these approaches are limited by their computational complexity. To reduce the computational cost of GCN, Hong et al. [41] proposed a minibatch GCN and aimed to utilize the advantages of CNN and GCN comprehensively. Furthermore, many superpixel-GCN methods have appeared. For instance, in Wan et al. [42], introduced superpixel into the GCN, which reduced the number of nodes and the amount of computation greatly. Subsequently, different superpixel-based graph neural networks (GNNs) were proposed. Ding et al. [43] proposed a multi-scale graph sample and aggregate network (GraphSAGE) for

HSI classification. Superpixel-based GCNs have achieve outstanding performance. However, the superpixel ignores the individual features of pixels. Specifically, the pixels in a superpixel are treated as the same label, which is not in line with the actual situation. Superpixel mechanism may cause incorrect classification of individual pixels in the superpixel.

To cope with the issues, a novel multi-feature fusion network, combing GCN and CNN, is proposed for HSI classification. In this network, superpixel-based GCN is proposed to refine the graph features. And the multi-scale graph mechanism is adopted to extract multi-scale spatial features from HSI. The multi-scale superpixel-based GCN can reduce the computing power requirements and deal with the problem of labeled deficiency. However, the local features for each pixel are ignored. Inspired by Hong et al. [41], the GCN and CNN are proposed to integrate into a single network. Concretely, a 3D CNN is decomposed into a 1D CNN and a 2D CNN. 1D CNN is used to extract spectral features from HSI for superpixel and subsequent multi-scale 2D CNN, which solves the problem of superpixel feature extraction and integrates the two networks of GCN and CNN. Then a dual CNN, i.e., 3×3 and 5×5 convolution kernel, is proposed to extract the multi-scale local features from HSI for pixels. Finally, the concatenate operation is introduced to fuse the multi-scale features of the four branches. In summary, the main contributions of the paper can be expressed as three folds:

- We propose a novel multi-scale fusion network by integrating multi-scale GCN with multi-scale CCN, which can extract multi-scale superpixel-based graph features and local pixel features.
- A spectral transformer mechanism is proposed, the spectral features of graph nodes can be extracted by 1D CNN. Then the noise of the original HSI is suppressed, and the adaptivity of GCN to different HSI is improved.
- A multi-scale neural network collaboration mechanism is explored in an end-to-end trainable network to predict the label of a pixel using multi-scale features.

3.2 Related work

In this part, some typical works, definitions, and notations (relate to the proposed work) about GCN on HSI classification are introduced.

3.2.1 Problem Definition

In this work, we attempt to achieve state-of-the-art classification accuracy for a large HSI cube with small labeled training samples by learning the relationship between labeled nodes and unlabeled nodes.

Problem Definition: Given a graph contains a node sets (superpixel sets), i.e., $\{(\boldsymbol{x}_i, \boldsymbol{y}_i)\}_{i=1}^{l}$ (labeled set), $\{\boldsymbol{x}_k\}_{k=l+1}^{l+u}$ (unlabeled set), and a label set $\mathcal{L} = \{1, .., c\}$ where $\{\boldsymbol{y}_i\}_{i=1}^{l} \in \mathcal{L}$, we seek to discover a suitable $f : \mathcal{X}^{l+u} \mapsto \mathcal{Y}^{l+u}$, which can utilize the relationship between $\{(\boldsymbol{x}_i, \boldsymbol{y}_i)\}_{i=1}^{l}$ and $\{\boldsymbol{x}_k\}_{k=l+1}^{l+u}$ and get a good performance.

3.2.2 Graph Convolution

Given a graph $\mathcal{G}$ with m nodes can be expressed as an adjacency matrix $A \in \{0,1\}^{m\times m}$ and an eigenmatrix $\boldsymbol{X} \in \mathbb{R}^{m\times d}$, that is, each node contains a B dimensional eigenvector. The mathematical expression of aggregation operation in GCN is as follows:

$$\boldsymbol{X}_{i+1} = \sigma\left(\boldsymbol{D}^{-\frac{1}{2}}\widehat{\boldsymbol{A}}\boldsymbol{D}^{-\frac{1}{2}}\boldsymbol{X}_i\boldsymbol{W}_i\right). \tag{3.1}$$

where, $\boldsymbol{X}_i$ represents the i-th layer output feature matrix and the $\boldsymbol{X}_0$ is set as $\boldsymbol{X}_0 = \boldsymbol{X}$. The node feature transformation of GCN from $\boldsymbol{X}_i \in \mathbb{R}^{m\times c_i}$ to $\boldsymbol{X}_{i+1} \in \mathbb{R}^{m\times c_{i+1}}$. $\widehat{\boldsymbol{A}} = \boldsymbol{A} + \boldsymbol{I}$, which means that the adjacency matrix is added with self-loop. $\boldsymbol{D}$ is the degree matrix, which performs normalization on $\widehat{\boldsymbol{A}}$. $\boldsymbol{W}_i \in \mathbb{R}^{c_i\times c_{i+1}}$ denotes a trainable weight matrix, which realizes the linear transformation on the feature matrix and σ is a nonlinear activation function, i.e., LeakyReLU.

3.3 Methodology

In this section, the brief framework of MFGCN is presented (Sect. 3.3.1). Then the pixel to region assignment and spectral feature refinement extraction are demonstrated (Sect. 3.3.2). Subsequently, the multi-scale GCN-based method and multi-scale CNN-based method are illustrated, respectively (Sects. 3.3.3 and 3.3.4). Afterward, region to pixel assignment and multi-feature fusion mechanism are introduced (Sect. 3.3.5). Finally, the details of the whole method process for HSI classification using MFGCN are shown (Sect. 3.3.6).

3.3.1 The Framework of Proposed MFGCN

In this part, the overview of MFGCN will be presented (see Fig. 3.1). Firstly, we adopt two-layer 1D CNN to extract spectral features of pixels in the original HSI. Simultaneously, the original HSI will be accurately partitioned into adaptive regions

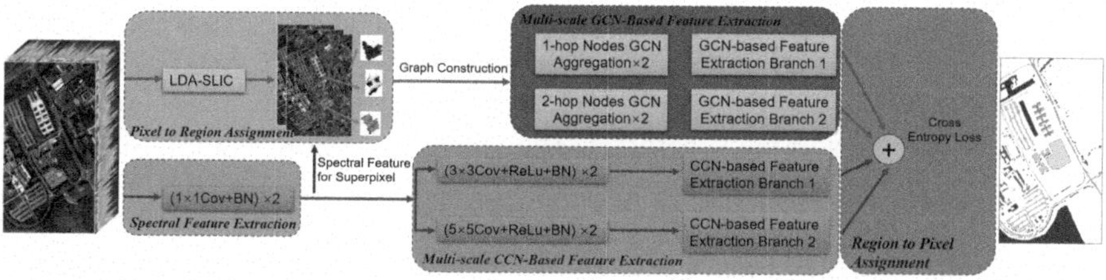

Fig. 3.1 The framework of MFGCN. The different colored blocks in the figure represent different modules in the model, and different colors of lines represent different information transmission. The proposed method mainly contains two branches, namely the multi-scale GCN-based method and the multi-scale CNN-based for HSI classification. Branch 1 adopts a semi-supervised classification method to solve the problem of labeled deficiency, and branch 2 utilizes CNN to address the problem of individual pixel classification errors

named superpixel by adopting LDA-SLIC, and the spectral features of each superpixel are calculated by employing a mean operation, which is extracted by 1D CNN. Subsequently, the graph is constructed in branch 1, and a multi-scale of GCN is proposed to extract the spectral-spatial feature of the graph. In branch 2, two 2D CNNs, i.e., 3×3 and 5×5 convolution kernel, are utilized to extract the spectral-spatial feature of HSI. Afterward, the concatenate operation is implemented to fussy the features extracted by different methods. Finally, the cross-entropy loss is adopted to interpret the features, and the label of each pixel can be predicted.

3.3.2 *Pixel to Region Assignment and Spectral Feature Transform*

The overview of the MFGCN is presented in Sect. 3.1. In this section, the HSI preprocessing of the proposed method will be demonstrated in detail.

(1) *Pixel to Region Assignment.* In the paper, linear discriminant analysis (LDA) [44] is adopted to reduce spectral dimensionality. Then the simple linear iterative clustering (SLIC) [45] is employed to segment the HSI into superpixels. Specifically, the LDA-SLIC method is adopted to segment the HSI into $F = \lceil (H \times W)/S \rceil$ superpixel, where H and W denote the height and the width of HSI, respectively. $S(1 \le S)$ is the segment scale of a superpixel, which controls the average pixels contained in the superpixel. That is, the number of superpixels is determined by S. Given an HSI cube $I_B = \{\boldsymbol{x}_1, \boldsymbol{x}_2, \cdots, \boldsymbol{x}_m\}$ with $m = H \times W$ pixels and B bands. We can express the superpixel as

$$\mathrm{HSI} = \cup_{i=1}^{F} S_i,\ \mathrm{S}_i \cap \mathrm{S}_j = \varnothing,\ i \neq j;\ i, j = 1,2, \ldots, S. \tag{3.2}$$

where S_i is a superpixel with n_i pixels; F denotes the number of superpixels.

(2) *Spectral Feature Transform.* In practice, it is a difficult task to extract the spectral features of superpixels. The general method is to extract the spectral value of each pixel from the original HSI directly, and then calculate the spectral average of pixels in the superpixel. The method is simple and intuitive, but it cannot suppress and eliminate the noise of original HSI by network training. To extract a discriminative and robust spectral feature for the proposed method, a novel two-layer 1D CNN is proposed. The output feature of the lth spectral convolutional layer in b th spectral channel $\mathbf{X}_b^l(\bullet)$ is expressed as

$$\boldsymbol{X}_b^l(\boldsymbol{p}_0) = \sigma\left(\boldsymbol{W}_b^l \cdot \tilde{\boldsymbol{X}}_b^{l-1}(\boldsymbol{p}_0) + a_b^l\right), \tag{3.3}$$

where $\boldsymbol{p}_0 = (x, y)$ is the spatial location of the pixel in HSI, $\boldsymbol{W}_b^l$ and a_b^l are trainable weight (1×1 convolutional kernel) and the bias, respectively. $\sigma(\bullet)$ presents the activate function, i.e., *ReLU*. As demonstrated in Eq. (3.3), the spectral feature vector of the pixel at the spatial location $\boldsymbol{p}_0$ in the proposed method can be written as

$$\boldsymbol{h}(\boldsymbol{p}_0) = (\boldsymbol{X}_1^2(\boldsymbol{p}_0), \boldsymbol{X}_2^2(\boldsymbol{p}_0), \cdots, \boldsymbol{X}_B^2(\boldsymbol{p}_0)), \tag{3.4}$$

Then, an association matrix $\boldsymbol{M} \in \mathbb{R}^{HW\times B}$ is proposed to present the relationship between pixels and superpixels, which construct a bridge between pixel and superpixel, and the GCN and the CNN are closely combined. Specifically, and the $\boldsymbol{M}$ can be calculated as

$$\boldsymbol{M}_{i,j} = \begin{cases} j \text{ if } \boldsymbol{x}_i \in S_j \\ 0 \text{ otherwise} \end{cases}, I_B = \text{Flatten }(HSI) \tag{3.5}$$

where $\boldsymbol{x}_i$ is the *ith* pixel of I_B, Flatten (HSI) denotes flattening operation in the spatial dimension for HSI. As described in Eq. (3.5), a good mapping between pixels and superpixels can be achieved.

Finally, the mean spectral feature of each superpixel is taken as a node vector, and the graph node can be mathematically expressed as

$$\begin{aligned} H &= [\boldsymbol{H}_1, \boldsymbol{H}_2, \ldots \boldsymbol{H}_F]^T \\ &= \left[\tfrac{1}{N_1}\textstyle\sum_{f=1}^{N_1} \boldsymbol{h}_k^1, \tfrac{1}{N_2}\sum_{f=1}^{N_2} \boldsymbol{h}_f^2, \ldots, \tfrac{1}{N_F}\sum_{f=1}^{N_F} \boldsymbol{h}_f^F\right]^T. \end{aligned} \tag{3.6}$$

where H_i is the ith node feature vector, and N_i denotes the number of pixels contained in superpixel, $\boldsymbol{h}_f^i$ is the spectral feature vector of the pixel shown as Eq. (3.4).

3.3.3 Multi-scale GCN-Based Method

The pixel to region assignment and spectral feature transform mechanisms are illustrated in detail in Sect. 3.3.2. However, the spatial relationships between nodes are lost, which is the difficulty should be confronted for graph construction. In order to solve the problem, reshape operation is adopted, namely

$$HSI = reshape\big(\boldsymbol{M}_{i,j}\boldsymbol{V}, \boldsymbol{H}\big), \tag{3.7}$$

where V_j in $\boldsymbol{V}$ denotes the centroid of the jth superpixel S_j, as long as we determine the position of V_j, the position of the superpixel is determined, that is, the position of V_j represents the position of the whole superpixel.

Through the reshape operation, we can project the graph features back to the spatial dimension. To improve computational efficiency, the superpixels are treated as graph nodes. Given a graph $\mathcal{G} = (\mathcal{V}, \mathcal{E}, \boldsymbol{A})$. the adjacency matrix $\boldsymbol{A}_{ij} \in \mathbb{R}^{N\times N}$ can be expressed as

$$\boldsymbol{A}_{ij} = \begin{cases} e^{-\gamma\|\boldsymbol{H}_i-\boldsymbol{H}_j\|^2}, & \text{if } \boldsymbol{x}_i \in \mathcal{N}_t\big(\boldsymbol{H}_j\big) \text{ or } \boldsymbol{x}_j \in \mathcal{N}_t(\boldsymbol{H}_i) \\ 0, & \text{otherwise} \end{cases}. \tag{3.8}$$

where $\boldsymbol{H}_i$ and $\boldsymbol{H}_j$ denote the spectral feature of nodes i, j, $\mathcal{N}_t\big(\boldsymbol{H}_j\big)$ is the t-hop neighbors of $\boldsymbol{H}_j$, $\gamma = 0.2$ is an empirical set. In our method, we can achieve the multi-scale operation by changing the aggregation different hop neighbors.

Then a multi-scale GCN network is proposed to process the graph. Referring Eq. (3.1), the lth layer output of graph convolution in branch 1 can be computed as

$$\boldsymbol{I}_1^l = LeakyReLU\left((\boldsymbol{D}_1)^{-\frac{1}{2}}\boldsymbol{A}_1(\boldsymbol{D}_1)^{-\frac{1}{2}}\boldsymbol{I}_1^{l-1}\boldsymbol{W}_1^l\right). \tag{3.9}$$

where $\boldsymbol{I}_1^{l-1}$ is the $l-1$ th layer output after normalization, $\boldsymbol{A}_1$ is the adjacency matrix aggregating 1-hop neighbors, $\boldsymbol{D}_1$ is the degree matrix of $\boldsymbol{A}_1$, $\boldsymbol{W}_1^l$ is a trainable weight matrix of lth layer in branch 1.

Similar to Eq. (3.9), the lth layer output of graph convolution in branch 2 can be expressed as

$$\boldsymbol{I}_2^l = LeakyReLU\left((\boldsymbol{D}_2)^{-\frac{1}{2}}\boldsymbol{A}_1(\boldsymbol{D}_2)^{-\frac{1}{2}}\boldsymbol{I}_2^{l-1}\boldsymbol{W}_2^l\right). \tag{3.10}$$

3.3.4 Multi-scale CNN-Based Method

3D CNN architecture possesses a good ability to extract local spectral-spatial information from HSI and has been widely adopted in HSI classification. However, 3D

CNN cannot integrate well with GCN. To solve the problem, we used the method of decomposing 3D CNN, namely, each 3D convolutional kernel is separated into a 2D convolutional kernel and a 1D convolutional kernel. Specifically, the 1D convolutional kernel is used to extract spectral features from HSI for superpixel and subsequent multi-scale 2D CNN. 2D CNN is used to extract spatial features from HSI, which solves the fusion problem of CNN and GCN, the parameters are reduced and robustness is enhanced (explained in Sect. 3.2) at the same time.

CNN with different convolution kernels have different receptive fields, so the local features extracted by the network are different. In practice, different classification targets require local features of different scales. In this work, a multi-scale CNN is proposed to extract the features, namely, two-branch network architecture with different kernels is adopted. Specifically, branch 1 adopts a 3 × 3 2D kernel, and branch 2 is a 5 × 5 2D kernel. The output of the lth convolutional layer in branch 1 can be calculated as

$$\boldsymbol{T}_{1j}^{l}(\boldsymbol{p}_0) = ReLU\left(\sum_{\boldsymbol{p}_n \in \mathcal{R}} \boldsymbol{k}_{1j}^{l}(\boldsymbol{p}_n) \cdot ReLU\left(\boldsymbol{W}_{1j}^{l} \cdot \boldsymbol{T}^{l-1}(\boldsymbol{p}_0 + \boldsymbol{p}_n)\right) + b_{1j}^{l}\right) \tag{3.11}$$

where $\boldsymbol{T}^{l-1}$ is the is the $l-1$th layer output after normalization. $\boldsymbol{k}_{1j}^{l}$, $\boldsymbol{W}_{1j}^{l}$ and b_{1j}^{l} denote jth 2D kernel, 1D kernel, and bias in branch 1. $\mathcal{R}$ defines a standard sampling grid, and $\boldsymbol{p}_n$ enumerates the locations.

3.3.5 Multi-feature Fusion and Region to Pixel Assignment

Four branches' networks are demonstrated in our method, i.e., two branches of GCNs and two branches of CNNs. Multi-scale GCN is used to solve the problem of labeled samples deficiency by semi-supervised learning. Simultaneously, multi-scale CNN can extract the multi-scale pixel-level feature to make up for the drawback of the superpixel-based method. Finally, the concatenate operation is adopted to fuse the features of the four branches, namely

$$\boldsymbol{Y} = Cat\left(\boldsymbol{I}_1^l, \boldsymbol{I}_2^l, \boldsymbol{T}_1^l, \boldsymbol{T}_2^l\right), \tag{3.12}$$

where $\boldsymbol{Y}$ is the output of multi-feature fusion, and $Cat(\bullet)$ denotes the concatenate operation.

Algorithm 1: MFGCN for HSI Classification.

Input: HSI; segment scale S ; learning rate γ ; number of epochs T.

Output: Pixel label.

The spectral feature is extracted by two layers 1×1 CNN referring Eq. (3.3). The whole image is segmented into superpixel via LDA-SLIC algorithm;

The superpixels input features are transformed by Eq. (3.5) and Eq. (3.6). The multi-scale graph is constructed by Eq. (3.7) and Eq. (3.8);

for t=1 to T do

  Extract multi-scale graph features by Eq. (3.9) and Eq. (3.10). Extract multi-scale pixel-wise feature by Eq. (3.11);

  Multi-feature fusion by Eq. (3.12) and pixel is classified by Eq. (3.13);

  We calculate the loss by Eq. (3.14) and adopt Adam gradient descent to updated the weight matrices;

end

Perform label prediction for each pixel.

Output: each pixel label.

To determine the label of each pixel, a softmax classifier is utilized to classify the output feature $\boldsymbol{O}$, namely

$$\boldsymbol{O}_i = \frac{e^{\boldsymbol{R}_i \cdot \boldsymbol{Y} + b_i}}{\sum_i^C e^{\boldsymbol{R}_i \cdot \boldsymbol{Y} + b_i}}, \tag{3.13}$$

where C presents the number of landcover categories, $\boldsymbol{R}_i$ and b_i denote the trainable parameter and bias.

3.3.6 HSI Classification With MFGCN

The entire process of feature extraction has been introduced in the previous article. Finally, the learnable parameters are penalized by the cross-entropy loss function, namely

$$\mathcal{L} = -\sum_{z \in \mathbf{y}_G} \sum_{f=1}^{C} \boldsymbol{Y}_{zf} \, ln \boldsymbol{O}_{Gzf}^{(final)}. \tag{3.14}$$

where $\boldsymbol{Y}_{zf}$ denotes the label matrix, $\mathbf{y}_G$ is the labeled example set, $\boldsymbol{O}_{Gzf}^{(final)}$ is the output of MFGCN. Meanwhile, we utilize the Adam [46] gradient descent to update the parameters of the proposed method. The implementation details of MFGN are demonstrated in Algorithm 1.

3.4 Experimental Results and Discussion

In this part, the performance of MFGCN is evaluated by extensive experimental comparison and analysis on three widely adopted datasets. Subsequently, the hyper-parameters selection is analyzed. Then, we study the ablative effects of GCN and CNN. Finally, the impact of limited training labeled samples on classification is investigated.

3.4.1 Dataset Description

(a) *Pavia University:* The first date set was captured by ROSIS over the University of Pavia, Italy, in 2003. The whole image contains 610 × 340 pixels and 103 spectral bands. The land covers in the dataset can be categorized into nine kinds of available inclusive classes. In the experiment, 30 pixels are randomly selected in each class from the HSI to train the models adopted in the paper. The detailed number of pixels is shown in Table 3.1. The false-color image, ground truth image, and reference map are shown in Fig. 3.2.

(b) *Salinas:* The second data set, Salinas, was collected by the AVIRIS sensor over Salinas Valley, California. The spatial resolution of the data is 3.7 m, and the size is 512 × 217. There are 224 spectral bands in the original data, and 204 bands are left after removing the water vapor absorption bands. The data contains 16 categories. In the experiment, 30 pixels are randomly selected in each class from the HSI to train the models adopted in the paper. The detailed number of pixels is shown in Table 3.2. The false-color image, ground truth image, and reference map are shown in Fig. 3.3.

(c) *Houston 2013:* The third data set was obtained by the ITRES CASI-1500 sensor and provided for the 2013 IEEE GRSS data fusion competition. The data size is 349 × 1905, including 144 bands ranging from 364 to 1046 nm. Ground cover is labeled into 15 categories. In the experiment, 30 pixels are randomly selected

Table 3.1 The detailed numbers of training and testing pixels of all classes in Pavia University data set

No. ID	Land cover name	Training	Testing
1	Asphalt	30	6601
2	Meadows	30	18619
3	Gravel	30	2069
4	Trees	30	3034
5	Metal sheets	30	1315
6	Bare soil	30	4999
7	Bitumen	30	1300
8	Self-blocking bricks	30	3652
9	Shadows	30	917

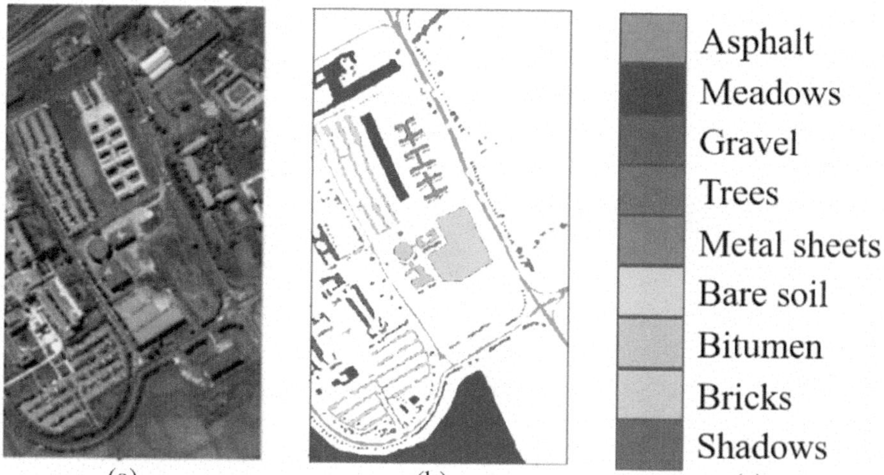

(a) (b) (c)

Fig. 3.2 **a** False-color image, **b** ground truth image, **c** reference map of Pavia University

Table 3.2 The detailed numbers of training and testing pixels of all classes in Salinas data set

No. ID	Land cover name	Training	Testing
1	Weed 1	30	1979
2	Weed 2	30	3696
3	Fallow	30	1946
4	Fallow plow	30	1364
5	Fallow smooth	30	2648
6	Stubble	30	3929
7	Celery	30	3549
8	Grapes untrained	30	11241
9	Soil	30	6137
10	Corn	30	3248
11	Lettuce 4 week	30	1038
12	Lettuce 5 week	30	1897
13	Lettuce 6 week	30	886
14	Lettuce 7 week	30	1040
15	Vineyard untrained	30	7238
16	Vineyard trellis	30	1777

in each class from the HSI to train the models adopted in the paper. The detailed number of pixels is shown in Table 3.3. The false-color composition picture, the corresponding ground truth map, and the reference map are demonstrated in Fig. 3.4.

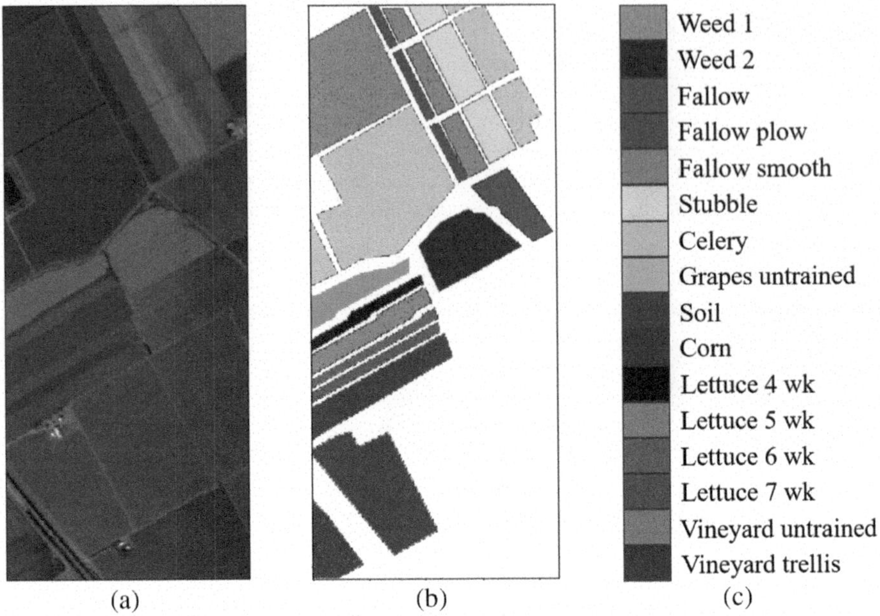

Fig. 3.3 **a** False-color image, **b** ground truth image, **c** reference map of Salinas

Table 3.3 The detailed numbers of training and testing pixels of all classes in Houston 2013 data set

No. ID	Land cover name	Training	Testing
1	Healthy grass	30	1221
2	Stressed grass	30	1224
3	Synthetic grass	30	667
4	Tress	30	1214
5	Soil	30	1212
6	Water	30	295
7	Residential	30	1238
8	Commercial	30	1214
9	Road	30	1222
10	Highway	30	1197
11	Railway	30	1205
12	Parking Lot 1	30	1203
13	Parking Lot 2	30	439
14	Tennis Court	30	398
15	Running Track	30	630

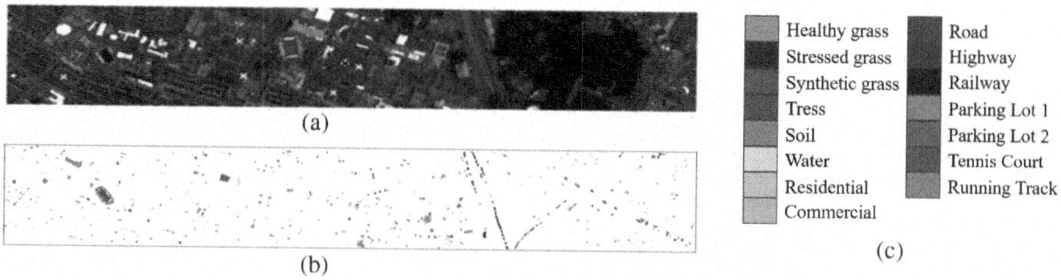

Fig. 3.4 **a** False-color image, **b** round truth image, **c** reference map of Houston 2013

3.4.2 Experimental Setup

To evaluate the superiority of proposed MFGCN, two traditional machine learning methods, i.e., multiband compact texture units (MBCTU) [47] and joint collaborative representation and SVM with decision fusion (JSDF) [48], two CNN-based methods, i.e., multi-scale dense network (MSDN) [49] and convolutional RNN (CRNN) [50], and two GNN methods, i.e., multi-scale graph sample and aggregate Network with context-Aware (MSAGE-CAL) [43] and spectral-spatial GCN (S^2GCN) [39], and are utilized as comparison models. The architecture of MFGCN is demonstrated in Table 3.4.

In MFGCN, three hyperparameters should be set, i.e., superpixel scale S, number of epochs T, and learning rate lr. The optimal hyperparameters are set as Table 3.5. The impacts of these parameters on classification are listed in detail in Sect. 3.4.4.

The experiments are carried out with a GeForce GTX 1080Ti 11G GPU and a 3.70G Intel i9-10900K CPU. And all experiments are run ten times, the standard deviation and the average are provided. In addition, to measure the classification accuracies of the comparison algorithms, the overall accuracy (OA), per-class accuracy (PA), average accuracy (AA), and kappa coefficient (Kappa) are adopted.

Table 3.4 The architectural details of MFGCN

Module	Details		
Pixel-to-region assignment	LDA SLIC	Spectral transform	(1×1 Cov + ReLU + BN) ×2
Graph construction	Calculate multi-scale adjacency matrix ***A***		
Multi-scale feature extraction	2×(GCN+LeakyReLU+BN) ×2		(3×3 Cov +ReLU+BN) ×2 (5×5 Cov +ReLU+BN) ×2
Original graph processing	Feature hybrid(concatenate)		
Output	Cross-entropy (target categories)		

Table 3.5 Hyperparameter settings for different datasets

Dataset	*S*	*T*	*lr*
Pavia University	100	600	0.001
Salinas	100	600	0.001
Houston 2013	100	600	0.001

3.4.3 Classification Results Discussion

(a) *Quantitative comparisons*

In this part, the MFGCN is quantitatively analyzed by four aforementioned indexes on three datasets to evaluate the performances. The quantitative results are shown in Tables 3.6, 3.7, and 3.8. The best results are reported in bold. From the results, it is obvious that the proposed MFGCN is superior to the comparative methods.

The classification accuracies achieved by seven methods on Pavia University are demonstrated by Table 3.6. From this table, we can observe that the CNN-based method and machine learning methods have achieved similar classification accuracy, MSDN and CRNN do not show their advantages on HSI classification compared with MBCUT and JSDF. This is because the CNNs are not sufficiently trained under the condition of limited training samples, and the deep features cannot be learned. Furthermore, GNN-based methods, i.e., MSAGE-CAL and MFGCN, achieve higher classification accuracies. However, S^2GCN is an exception. This is mainly because MSAGE-CAL and MFGCN adopt superpixel and multi-scale mechanisms, which improve the methods to learn more spatial features from HSI. It is also should be notable that, compared with MSAGE-CAL, MFGCN improves the OA, AA, and Kappa by 3.35, 4.78, and 2.19%, respectively. This is because the CNN, adopted in MFGCN, can learn the local features of pixels.

As for the Salinas dataset, in Table 3.7, the MFGCN outperforms the competitive methods. Specifically, the performance of OA, AA, and Kappa, achieved by MFGCN, are improved by 2.41, 2.69, and 2.14%. Meanwhile, we can observe that MFGCN achieves the highest PA in most classes, especially in C8 (Grapes untrained) and C15 (Vineyard untrained). The results demonstrate the proposed method possesses good classification adaptability to different objects with similar spectrum. Furthermore, JSDF gets an outstanding result, which illustrates that traditional machine learning methods can also achieve results comparable to deep learning with proper design.

Table 3.8 presents the experimental results on Houston 2013. Compared with the Pavia University and Salinas datasets, Houston 2013 dataset is larger and contains more details to be classified, which are the challenge to classifiers. From the results, we can observe that the MFGCN has a good detail classification effect, and the PAs in most classes are highest. In addition, compared with MSAGE-CAL, the OA, AA, and kappa are improved by 3.11, 3.36, and 2.38%, respectively. The results show the design of the proposed network is reasonable, and the multi-scale CNNs can effectively extract the features of the individual pixels.

Table 3.6 Quantitative experimental performance on Pavia University, bold is the best performance

Class No.	MSDN	CRNN	S^2GCN	MSAGE-CAL	MBCUT	JSDF	MFGCN
1	91.20±1.62	79.85±4.62	92.87±3.79	93.93±1.02	87.49±3.99	82.40±4.07	**99.78±0.20**
2	94.61±0.59	82.33±3.17	87.06±4.47	99.90±0.10	89.11±5.58	90.76±3.74	**99.99±0.01**
3	87.18±2.33	89.67±3.69	87.97±4.77	89.75±2.46	86.24±4.23	86.71±4.14	**98.86±1.20**
4	90.29±3.21	91.45±2.44	90.85±0.94	92.16±0.82	90.61±3.39	92.88±2.16	**95.16±1.70**
5	95.64±0.76	94.12±1.78	**100.00±0.00**	98.71±1.29	97.18±1.18	**100.00±0.00**	99.94±0.12
6	88.89±1.28	91.37±2.11	88.69±2.64	82.88±3.46	93.25±2.93	94.30±4.55	**99.96±0.04**
7	92.48±2.36	93.67±1.97	98.88±1.08	99.54±0.26	93.49±2.47	96.62±1.37	**99.23±0.15**
8	82.67±1.34	80.18±3.29	89.97±3.28	96.55±1.71	84.14±4.78	94.69±3.74	**99.35±0.59**
9	95.17±0.57	82.34±4.86	98.89±0.53	96.40±1.33	96.57±1.22	99.56±0.36	**99.83±0.13**
OA (%)	90.90±1.49	85.46±1.75	89.74±1.70	96.14±1.10	89.43±2.14	90.82±1.30	**99.49±0.16**
AA (%)	91.12±1.56	87.22±1.82	92.80±0.47	94.42±0.87	90.90±0.89	93.10±0.65	**99.20±0.28**
Kappa	91.55±1.63	0.8419±1.43	86.65±2.06	97.12±1.49	86.24±2.62	88.02±1.62	**99.32±0.21**

Table 3.7 Quantitative experimental performance on Salinas, bold is the best performance

Class No.	MSDN	CRNN	S^2GCN	MSAGE-CAL	MBCUT	JSDF	MFGCN
1	98.21±0.67	99.34±0.53	99.01±0.44	**100.00±0.00**	99.18±0.80	**100.00±0.00**	99.98±0.02
2	99.62±0.19	99.17±0.21	99.18±0.59	99.95±0.05	99.76±0.33	**100.00±0.00**	**100.00±0.00**
3	97.43±0.52	96.54±1.39	97.15±2.76	99.90±0.10	99.13±1.04	**100.00±0.00**	**100.00±0.00**
4	97.62±0.97	97.32±0.27	99.11±0.55	**99.93±0.07**	97.61±0.82	**99.93±0.09**	99.76±0.29
5	95.31±1.21	98.75±0.89	97.55±2.35	85.33±4.39	96.54±1.01	**99.77±0.31**	99.12±0.35
6	99.44±0.26	99.19±0.77	99.32±0.35	98.98±1.02	99.74±0.32	**100.00±0.00**	**100.00±0.00**
7	97.28±0.61	98.67±1.30	90.06±0.27	**100.00±0.00**	98.26±1.64	99.99±0.01	99.97±0.03
8	77.21±2.37	72.38±3.98	70.68±5.20	92.59±2.17	81.98±4.32	87.79±4.89	**98.40±1.70**
9	98.62±0.51	97.29±0.61	98.32±1.79	99.98±0.02	99.47±0.51	99.67±0.33	**100.00±0.00**
10	93.18±1.39	91.44±1.64	90.97±2.59	97.27±2.73	92.21±2.75	96.53±2.55	**97.40±2.21**
11	96.21±0.39	96.82±0.78	98.00±1.65	96.47±0.92	96.24±2.68	99.76±0.21	**99.87±0.12**
12	99.99±0.01	99.21±0.32	99.56±0.59	99.74±0.26	98.98±0.45	**100.00±0.00**	**100.00±0.00**
13	96.49±0.92	97.29±0.86	97.83±0.72	97.32±2.68	96.73±1.66	**100.00±0.00**	99.94±0.11
14	95.21±0.79	95.10±1.73	95.75±1.65	93.62±3.17	96.50±3.05	98.71±0.72	**99.05±0.43**
15	73.62±2.31	76.33±4.62	70.36±3.62	90.69±2.81	79.41±5.67	81.86±5.26	**99.01±0.78**
16	96.31±0.52	97.99±0.61	96.90±1.97	97.15±1.23	96.89±2.19	98.99±0.63	**99.58±0.57**
OA (%)	91.64±1.67	87.64±0.83	88.39±1.01	96.87±1.20	92.14±0.86	94.67±0.77	**99.28±0.37**
AA (%)	94.48±0.85	94.55±0.57	94.30±0.47	96.81±0.88	95.54±0.56	97.69±0.34	**99.50±0.17**
Kappa	90.62±0.89	86.72±1.01	87.10±1.12	97.06±0.92	91.25±0.95	94.06±0.85	**99.20±0.41**

Table 3.8 Quantitative experimental performance on Houston 2013, bold is the best performance

Class No.	MSDN	CRNN	S^2GCN	MSAGE-CAL	MBCUT	JSDF	MFGCN
1	92.62±1.25	82.45±2.19	96.30±3.07	87.21±2.31	92.86±3.83	**97.41±1.21**	96.03±5.07
2	95.28±0.61	84.12±3.64	98.57±1.47	93.74±1.26	92.18±2.79	**99.84±0.25**	98.61±1.12
3	92.33±0.79	91.56±1.07	98.88±0.43	97.01±0.92	97.42±1.19	99.88±0.22	**100.00±0.00**
4	93.46±1.35	91.29±3.78	97.68±2.89	95.21±1.20	90.96±1.98	98.22±2.80	**99.14±1.90**
5	99.21±0.09	98.81±0.62	97.66±1.12	98.94±1.06	97.17±1.29	**100.00±0.00**	**100.00±0.00**
6	88.92±1.33	94.83±2.19	96.84±1.17	94.03±1.47	91.78±3.22	**99.32±1.09**	91.64±6.77
7	85.21±0.92	86.42±3.42	83.48±5.89	91.35±2.91	82.88±3.81	91.32±4.91	**97.64±1.31**
8	79.64±3.66	53.05±8.71	76.15±4.37	82.62±3.61	71.85±5.64	68.82±6.16	**85.42±2.97**
9	81.21±2.39	84.04±4.63	82.17±1.78	88.31±1.74	81.94±4.25	69.47±8.56	**94.29±1.97**
10	78.62±3.72	45.14±8.77	86.85±8.32	96.75±1.35	87.31±5.08	85.63±9.32	**98.70±1.90**
11	86.42±1.51	61.85±9.39	88.57±5.06	92.19±0.91	77.41±6.46	94.51±3.82	**96.80±3.53**
12	81.92±1.37	84.40±2.93	78.64±4.79	90.32±2.68	86.35±5.85	84.33±5.33	**93.14±4.57**
13	83.27±0.68	84.14±3.12	75.62±6.93	88.44±3.76	85.58±5.35	98.10±1.28	**94.32±7.85**
14	97.77±0.68	96.03±0.73	99.45±0.44	98.28±0.42	96.85±1.85	**100.00±0.00**	**100.00±0.00**
15	97.65±0.24	93.45±2.41	98.03±1.07	98.36±0.94	92.27±3.32	**99.86±0.36**	97.46±1.53
OA (%)	88.49±0.96	82.10±1.21	89.31±1.00	92.13±1.02	87.07±1.12	90.51±0.95	**95.24±0.51**
AA (%)	88.90±1.37	79.21±1.02	90.33±1.06	92.85±0.76	89.32±1.08	92.46±0.75	**96.21±0.37**
Kappa	88.27±1.65	77.61±1.19	88.44±1.08	92.47±1.23	86.01±1.21	89.74±1.03	**94.85±0.55**

(b) *Qualitative comparisons*

The classification maps acquired by seven methods on the three datasets are demonstrated in Figs. 3.5, 3.6, and 3.7, respectively. In the classification maps, different land-cover classes are presented in different colors. When comparing the sub-figures with the ground-truth maps on three datasets, we can observe that the MFGCN achieves more reasonable maps than the comparative methods. Specifically, the proposed method can refine the local features of the pixel and has a good classification effect on the details. For example, the result maps contain less miss-classification on Pavia University and Houston 2013. Moreover, a spectral extraction network is proposed by the MFGCN to transform the spectral feature extraction for superpixels, which improves the ability of the algorithm to suppress spectral noise. As shown in Fig. 3.5, it is clear that the MFGCN demonstrates excellent classification abilities in C8 (Grapes-untrained) and C15 (Vineyard-untrained). In summary, the visual results demonstrate that our network design is reasonable for HSI classification.

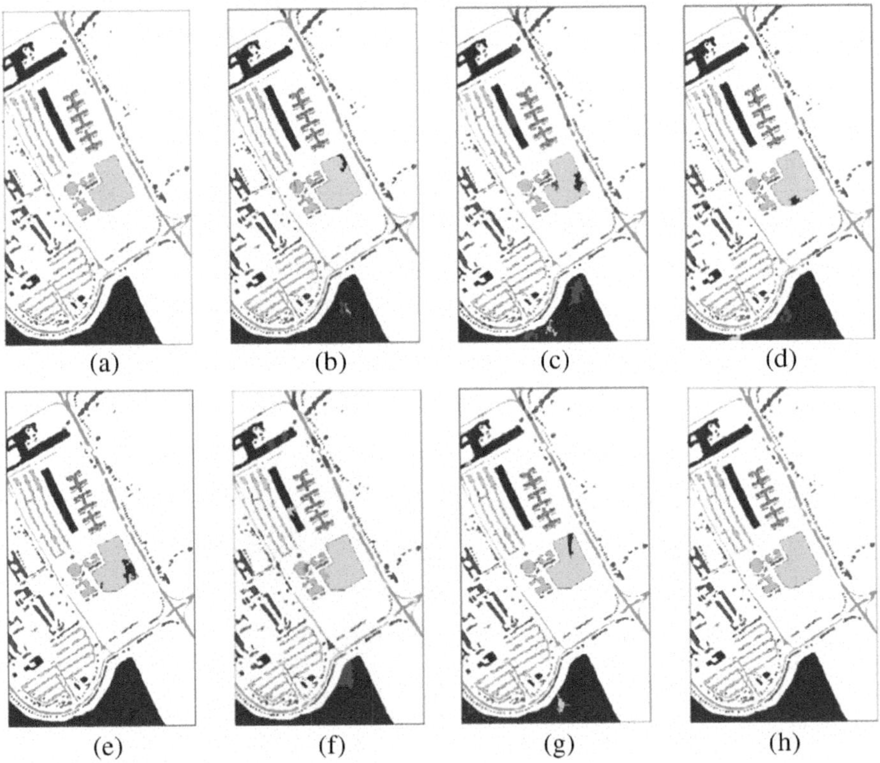

Fig. 3.5 Visual diagram of classification results on Pavia University dataset. **a** Ground truth; **b** MSDN; **c** CRNN; **d** S^2GCN; **e** MSAGE-CAL; **f** MBCUT; **g** JSDF; **h** MFGCN

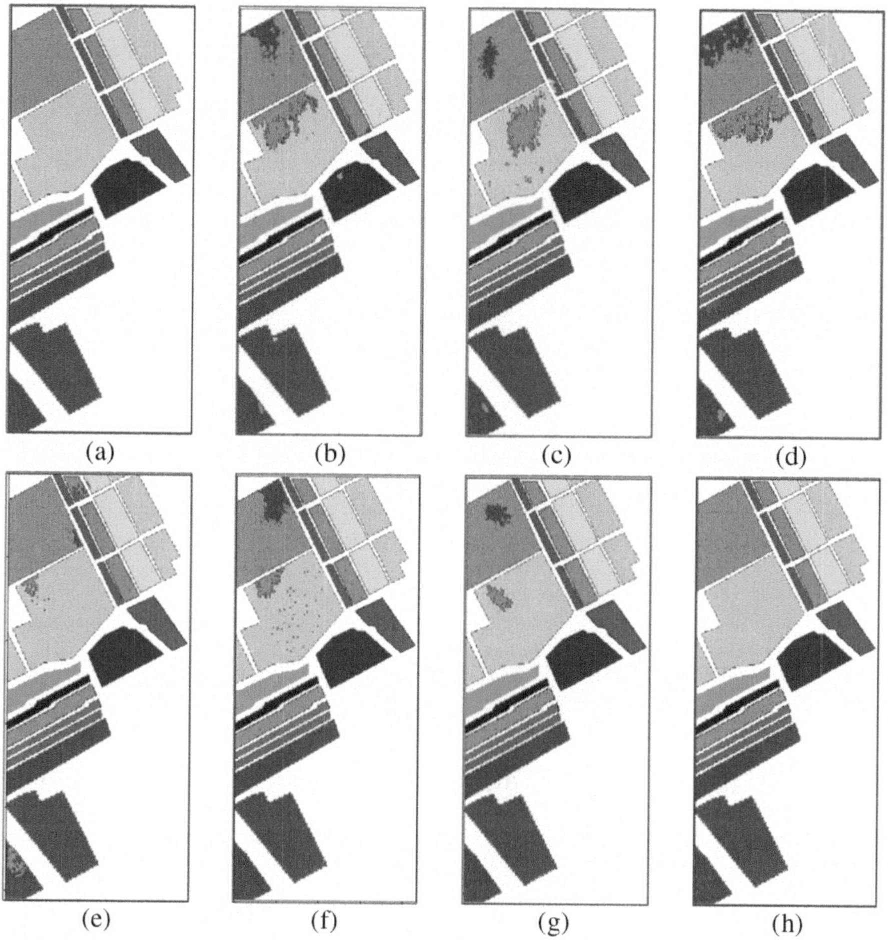

Fig. 3.6 Visual diagram of classification results on Salinas dataset. **a** Ground truth; **b** MSDN; **c** CRNN; **d** S2GCN; **e** MSAGE-CAL; **f** MBCUT; **g** JSDF; **h** MFGCN

3.4.4 Hyperparameter Selection

Superpixel scale S, number of epochs T, and learning rate lr have important impacts on the performance of the proposed model. The hyperparameter settings are shown in Table 3.5. In this experiment, a grid search strategy is utilized to find the optimal setting. And OA is adopted to report the performance of proposed methods with different parameter settings.

(a) *Effect of lr and T selection*

In the experiment, the sensitivity of lr and T is investigated in detail. To analyze the impacts of two parameters on the model, the S is fixed as 100. The lr is set as

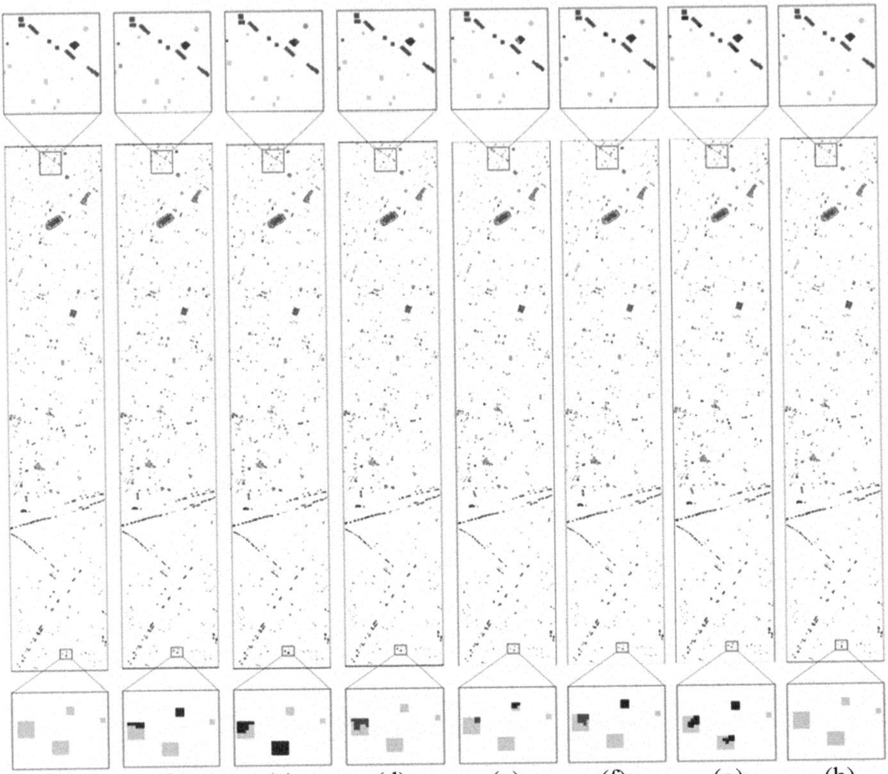

Fig. 3.7 Visual diagram of classification results on Houston 2013 dataset. **a** Ground truth; **b** MSDN; **c** CRNN; **d** S^2 GCN; **e** MSAGE-CAL; **f** MBCUT; **g** JSDF; **h** MFGCN

0.1, 0.01, 0.001, 0.001. And the T is varied from 200 to 1000 with an interval of 200. Fig. 3.8 shows the OA value under different parameter combinations on three datasets. From the results, we can observe that the proposed method achieves the best accuracy as T is set as 600 on Pavia University. As for the Salinas dataset and Houston 2013, setting the T to 600–800 is the best choice. However, the larger T need more time to train the model. In terms of lr, a large lr can speed up the model parameter training, however we cannot get the optimal parameter. We can archive a good OA with a smaller lr, however more time is required to train the model. Considering classification accuracy and learning efficiency, the lr in our proposed method is set as 0.001.

(b) *Effect of S selection*

The superpixel scale S determines the size of the superpixels and affects the classification accuracy of MFGCN. In the experiment, the S is varied from 100 to 500 with an interval of 100. The experiment setups are the same as Sect. 3.2, and the results are reported in Fig. 3.9. As seen, the classification accuracies of the proposed method

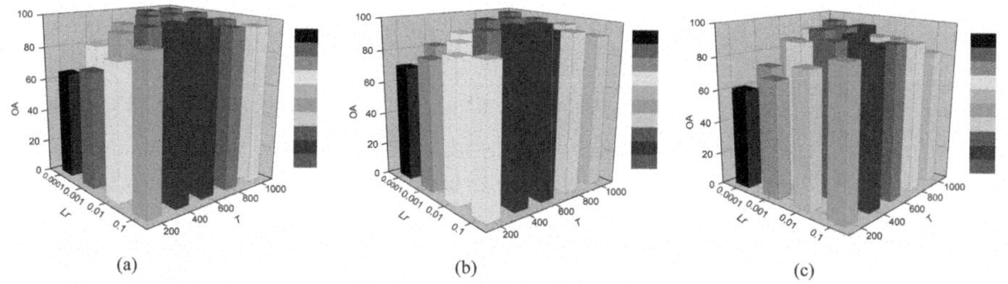

Fig. 3.8 Effect of *lr* and *T* for MFGCN. **a** Pavia University. **b** Salinas. **c** Houston 2013

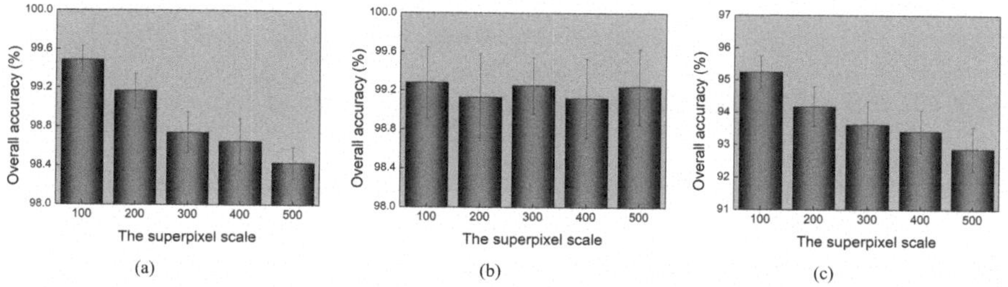

Fig. 3.9 Effect of *S* for MFGCN. **a** Pavia University. **b** Salinas dataset. **c** Houston 2013 dataset

on three datasets decrease with the increase of S. This is mainly because a smaller S leads to a smaller superpixel that can preserve more local features. A smaller S can improve the classification accuracy while including more superpixels. That is to say, a smaller S causes a larger graph, and more computational ability is needed to train the model. Due to limitations of the experimental computer, the S is fixed to 100 in our proposed method.

3.4.5 Ablation Study

As introduced in the paper, the proposed MFGCN is mainly composed of two parts, i.e., multi-sale GCN network, and multi-sale CNN network. And spectral transform mechanism, multi-scale operations, GCN networks and CNN networks are critical important to improving the classification performance. In the experiment, the Pavia University dataset is utilized to evaluate the contributions of these operations. And OA, AA, and Kappa are adopted to report the results produced by MFGCN without one of the four operations. For simplicity, we utilize "MFGCN-v_1", "MFGCN-v_2", "MFGCN-v_3" and "MFGCN-v_4" to represent the reduced methods that remove spectral transform mechanism, multi-scale operations, GCN networks, and CNN networks, respectively. The comparative results are exhibited in Table 3.9. The best results are marked in bold. From the results, the MFGCN achieves higher OA AA

Table 3.9 Pavia University ablation study, best performance in bold

ID.	MFGCN -v_1	MFGCN -v_2	MFGCN -v_3	MFGCN -v_4	MFGCN
OA (%)	97.39±0.19	98.16±0.52	88.83±0.23	94.27±0.51	**99.49±0.16**
AA (%)	96.82±0.25	98.47±0.47	89.13±0.18	93.59±0.67	**99.20±0.28**
Kappa	97.16±0.29	98.31±0.29	88.47±0.32	94.18±0.29	**99.32±0.21**

and Kappa over MFGCN-v_1, MFGCN-v_2, MFGCN-v_3, and MFGCN-v_4, which verifies the contributions of spectral transform mechanism, multi-scale operations, GCN networks, and CNN networks to performance improvement.

3.4.6 Limited Labeled Samples for Classification

Due to the labeled pixel being expensive, HSI confronts the problem of labeled samples deficiency. In this experiment, the classification ability of MFGCN with limited training samples is analyzed. Specifically, the number of training pixels is varied from 5 to 30 with an interval of 5. The OA is adopted to evaluate the performance of all comparison methods. And the experimental setups are the same as Sect. 3.4.2. The results of all investigated methods on three datasets are demonstrated in Fig. 3.10. As shown in Fig. 3.10, we can observe that the performances of the proposed method surpass all competitive models on three datasets. Concretely, the OA performances of all investigated methods are improved as the training data increases. Because GCN can learn the relationships between neighbor nodes, GCN-based methods outperform the other methods. In addition, the performance of the CNN-based method is unstable, especially when the number of labeled data is small. Since our proposed model comprehensively utilizes the advantages of CNN and GCN which proves superior adaptability with limited samples.

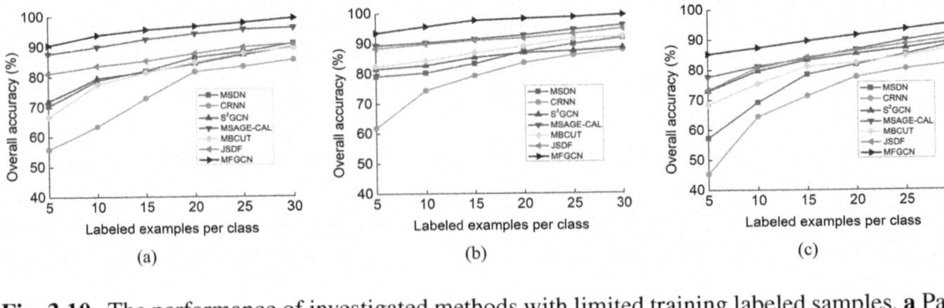

Fig. 3.10 The performance of investigated methods with limited training labeled samples. **a** Pavia University; **b** Salinas; **c** Houston 2013

3.5 Conclusion

In this paper, we introduce a novel HSI classification method combing multi-scale GCN and multi-scale CNN. To learn the semantic structure relationships of land covers and reduce the computational complexity, a superpixel method is proposed. In our proposed method, a 1D CNN is adopted to extract the spectral features for superpixels (nodes), which is different from most existing methods. Then a graph is constructed, and two branches GCNs are utilized to extract the multi-scale features from the graph. To extract the local features of the pixel and classify HSI at the pixel level, a multi-scale CNN network is proposed to extract the multi-scale features based on pixel level. Finally, a concatenate operation is employed to fuse the complementary multi-scale features. In order to evaluate the performance of the proposed model, extensive experiments have been implemented. In comparison with the state-of-the-art models on three datasets, the proposed method achieves superior experimental results and outperforms competitive methods.

References

1. Y. Ding, X. Zhao, Z. Zhang, W. Cai, N. Yang, Graph sample and aggregate-attention network for hyperspectral image classification. IEEE Geosci. Remote Sens. Lett. **19**, 1–5 (2022)
2. D. Hong, W. He, N. Yokoya, J. Yao, L. Gao, L. Zhang, J. Chanussot, X. Zhu, Interpretable hyperspectral artificial intelligence: when nonconvex modeling meets hyperspectral remote sensing. IEEE Geosci. Remote Sens. Magaz. **9**(2), 52–87 (2021)
3. D. Yao, Z. Zhi-li, Z. Xiao-feng, C. Wei, H. Fang, C. Yao-ming, Deep hybrid: multi-graph neural network collaboration for hyperspectral image classification. Def. Technol. (2022)
4. D. Hong, L. Gao, N. Yokoya, J. Yao, J. Chanussot, Q. Du, B. Zhang, More diverse means better: multimodal deep learning meets remote-sensing imagery classification. IEEE Trans. Geosci. Remote Sens. **59**(5), 4340–4354 (2020)
5. P. Zhong, Z. Gong, J. Shan, Multiple instance learning for multiple diverse hyperspectral target characterizations. IEEE Trans. Neural Netw. Learn. Syst. **31**(1), 246–258 (2019)
6. S. Li, W. Song, L. Fang, Y. Chen, P. Ghamisi, J.A. Benediktsson, Deep learning for hyperspectral image classification: an overview. IEEE Trans. Geosci. Remote Sens. **57**(9), 6690–6709 (2019)
7. D. Hong, N. Yokoya, J. Chanussot, X.X. Zhu, CoSpace: Common subspace learning from hyperspectral-multispectral correspondences. IEEE Trans. Geosci. Remote Sens. **57**(7), 4349–4359 (2019)
8. Z.Z.Y. Ding, X. Zhao, D. Hong, W. Li, W. Cai, Y. Zhan, AF2GNN: graph convolution with adaptive filters and aggregator fusion for hyperspectral image classification. Inform. Sci. (2022)
9. Y. Cai, M. Zeng, Z. Cai, X. Liu, Z. Zhang, Graph regularized residual subspace clustering network for hyperspectral image clustering. Inform. Sci. **578**, 85–101 (2021)
10. Y. Ding, X. Zhao, Z. Zhang, W. Cai, N. Yang, Y. Zhan, Semi-supervised locality preserving dense graph neural network with ARMA filters and context-aware learning for hyperspectral image classification. IEEE Trans. Geosci. Remote Sens. (2021)
11. J. Bai, A. Yuan, Z. Xiao, H. Zhou, D. Wang, H. Jiang, L. Jiao, Class incremental learning with few-shots based on linear programming for hyperspectral image classification. IEEE Trans. Cybern. (2020)
12. J. Peng, L. Li, Y.Y. Tang, Maximum likelihood estimation-based joint sparse representation for the classification of hyperspectral remote sensing images. IEEE Trans. Neural Netw. Learn. Syst. **30**(6), 1790–1802 (2018)

13. Y. Cai, Z. Zhang, Z. Cai, X. Liu, X. Jiang, Q. Yan, Graph convolutional subspace clustering: a robust subspace clustering framework for hyperspectral image. IEEE Trans. Geosci. Remote Sens. **59**(5), 4191–4202 (2020)
14. J. Bai, S. Huang, Z. Xiao, X. Li, Y. Zhu, A.C. Regan, L. Jiao, Few-shot hyperspectral image classification based on adaptive subspaces and feature transformation. IEEE Trans. Geosci. Remote Sens. **60**, 1–17 (2022)
15. C. Chen, W. Li, H. Su, K. Liu, Spectral-spatial classification of hyperspectral image based on kernel extreme learning machine. Remote Sens. **6**(6), 5795–5814 (2014)
16. J. Li, J.M. Bioucas-Dias, A. Plaza, Semisupervised hyperspectral image segmentation using multinomial logistic regression with active learning. IEEE Trans. Geosci. Remote Sens. **48**(11), 4085–4098 (2010)
17. C. Bo, H. Lu, D. Wang, Hyperspectral image classification via JCR and SVM models with decision fusion. IEEE Geosci. Remote Sens. Lett. **13**(2), 177–181 (2015)
18. Y. Cai, X. Liu, Z. Cai, BS-Nets: an end-to-end framework for band selection of hyperspectral image. IEEE Trans. Geosci. Remote Sens. **58**(3), 1969–1984 (2019)
19. M. Fauvel, J.A. Benediktsson, J. Chanussot, J.R. Sveinsson, Spectral and spatial classification of hyperspectral data using SVMs and morphological profiles. IEEE Trans. Geosci. Remote Sens. **46**(11), 3804–3814 (2008)
20. L. Fang, N. He, S. Li, P. Ghamisi, J.A. Benediktsson, Extinction profiles fusion for hyperspectral images classification. IEEE Trans. Geosci. Remote Sens. **56**(3), 1803–1815 (2017)
21. S. Jia, L. Shen, Q. Li, Gabor feature-based collaborative representation for hyperspectral imagery classification. IEEE Trans. Geosci. Remote Sens. **53**(2), 1118–1129 (2014)
22. Z. Tang, M. Ling, H. Yao, Z. Qian, X. Zhang, J. Zhang, S. Xu, Robust image hashing via random Gabor filtering and DWT. Comput. Mater. Cont. **55**(2), 331–344 (2018)
23. P. Quesada-Barriuso, F. Argüello, D.B. Heras, Spectral–spatial classification of hyperspectral images using wavelets and extended morphological profiles. IEEE J. Select. Top. Appl. Earth Observ. Remote Sens. **7**(4), 1177–1185 (2014)
24. J.A. Benediktsson, J.A. Palmason, J.R. Sveinsson, Classification of hyperspectral data from urban areas based on extended morphological profiles. IEEE Trans. Geosci. Remote Sens. **43**(3), 480–491 (2005)
25. X. Kang, S. Li, J.A. Benediktsson, Spectral–spatial hyperspectral image classification with edge-preserving filtering. IEEE Trans. Geosci. Remote Sens. **52**(5), 2666–2677 (2013)
26. X.X. Zhu, D. Tuia, L. Mou, G.-S. Xia, L. Zhang, F. Xu, F. Fraundorfer, Deep learning in remote sensing: a comprehensive review and list of resources. IEEE Geosci. Remote Sens. Magaz. **5**(4), 8–36 (2017)
27. L. Mou, P. Ghamisi, X.X. Zhu, Deep recurrent neural networks for hyperspectral image classification. IEEE Trans. Geosci. Remote Sens. **55**(7), 3639–3655 (2017)
28. Y. Chen, Z. Lin, X. Zhao, G. Wang, Y. Gu, Deep learning-based classification of hyperspectral data. IEEE J. Select. Top. Appl. Earth Observ. Remote Sens. **7**(6), 2094–2107 (2014)
29. K. Makantasis, K. Karantzalos, A. Doulamis, N. Doulamis, Deep supervised learning for hyperspectral data classification through convolutional neural networks, in *Proceedings of the 2015 IEEE International Geoscience and Remote Sensing Symposium (IGARSS)* (IEEE, 2015), pp. 4959–4962
30. S.K. Roy, G. Krishna, S.R. Dubey, B.B. Chaudhuri, HybridSN: exploring 3-D–2-D CNN feature hierarchy for hyperspectral image classification. IEEE Geosci. Remote Sens. Lett. **17**(2), 277–281 (2019)
31. W. Hu, Y.Y. Huang, L. Wei, F. Zhang, H.C. Li, Deep convolutional neural networks for hyperspectral image classification. J. Sens. **2015**, 1–12 (2015)
32. Y. Li, H. Zhang, Q. Shen, Spectral–spatial classification of hyperspectral imagery with 3D convolutional neural network. Remote Sens. **9**(1), 67 (2017)
33. J. Yang, Y.-Q. Zhao, J.C.-W. Chan, Learning and transferring deep joint spectral–spatial features for hyperspectral classification. IEEE Trans. Geosci. Remote Sens. **55**(8), 4729–4742 (2017)
34. C. Chen, J.-J. Zhang, C.-H. Zheng, Q. Yan, L.-N. Xun, Classification of hyperspectral data using a multi-channel convolutional neural network, in *International Conference on Intelligent Computing* (Springer, 2018), pp. 81–92

35. J. Long, E. Shelhamer, T. Darrell, Fully convolutional networks for semantic segmentation, in *Proceedings of the IEEE Conference on Computer Vision and Pattern Recognition* (2015), pp. 3431–3440
36. L. Zhu, Y. Chen, P. Ghamisi, J.A. Benediktsson, Generative adversarial networks for hyperspectral image classification. IEEE Trans. Geosci. Remote Sens. **56**(9), 5046–5063 (2018)
37. H. Hu, M. Yao, F. He, F. Zhang, Graph neural network via edge convolution for hyperspectral image classification. IEEE Geosci. Remote Sens. Lett. **19**, 1–5 (2022)
38. M.W.T.N. Kipf, *Semi-Supervised Classification with Graph Convolutional Networks* (2016)
39. A. Qin, Z. Shang, J. Tian, Y. Wang, T. Zhang, Y.Y. Tang, Spectral–spatial graph convolutional networks for semisupervised hyperspectral image classification. IEEE Geosci. Remote Sens. Lett. **16**(2), 241–245 (2018)
40. J. Bai, B. Ding, Z. Xiao, L. Jiao, H. Chen, A.C. Regan, Hyperspectral image classification based on deep attention graph convolutional network. IEEE Trans. Geosci. Remote Sens. **60**, 1–16 (2021)
41. D. Hong, L. Gao, J. Yao, B. Zhang, A. Plaza, J. Chanussot, Graph convolutional networks for hyperspectral image classification. IEEE Trans. Geosci. Remote Sens. **59**(7), 5966–5978 (2020)
42. S. Wan, C. Gong, P. Zhong, B. Du, L. Zhang, J. Yang, Multiscale dynamic graph convolutional network for hyperspectral image classification. IEEE Trans. Geosci. Remote Sens. **58**(5), 3162–3177 (2019)
43. Y. Ding, X. Zhao, Z. Zhang, W. Cai, N. Yang, Multiscale graph sample and aggregate network with context-aware learning for hyperspectral image classification. IEEE J. Select. Top. Appl. Earth Observ. Remote Sens. **14**, 4561–4572 (2021)
44. A.J. Izenman, Linear discriminant analysis, in *Modern Multivariate Statistical Techniques* (Springer, 2013), pp. 237–280
45. R. Achanta, A. Shaji, K. Smith, A. Lucchi, P. Fua, S. Süsstrunk, SLIC superpixels compared to state-of-the-art superpixel methods. IEEE Trans. Pattern Anal. Mach. Intell. **34**(11), 2274–2282 (2012)
46. T. Dozat, *Incorporating Nesterov Momentum into Adam* (2016)
47. K. Djerriri, A. Safia, R. Adjoudj, M.S. Karoui, Improving hyperspectral image classification by combining spectral and multiband compact texture features, in *IGARSS 2019–2019 IEEE International Geoscience and Remote Sensing Symposium* (IEEE, 2019), pp. 465–468
48. W. Li, G. Wu, F. Zhang, Q. Du, Hyperspectral image classification using deep pixel-pair features. IEEE Trans. Geosci. Remote Sens. **55**(2), 844–853 (2016)
49. C. Zhang, G. Li, S. Du, Multi-scale dense networks for hyperspectral remote sensing image classification. IEEE Trans. Geosci. Remote Sens. **57**(11), 9201–9222 (2019)
50. M. Zhang, W. Li, Q. Du, Diverse region-based CNN for hyperspectral image classification. IEEE Trans. Image Process. **27**(6), 2623–2634 (2018)

Chapter 4
Multi-receptive Field: An Adaptive Path Aggregation Graph Neural Framework for Hyperspectral Image Classification

4.1 Introduction

Hyperspectral images (HSIs) collect rich spatial-spectral information in hundreds of spectral bands, which are captured by hyperspectral remote sensors, [1–4], which are more effective in distinguishing different land-covers compared with other multispectral [5] or RGB (red, green, and blue) image [6]. Therefore, HSIs are widely employed in various applications, ranging from military reconnaissance, marine monitoring to disaster prevention and control [7, 8]. HSI classification (category each pixel into certain label) is a crucial technique for these applications. However, the complex noise effects and spectral variability [9], labeled training samples deficiency [10], and high spectral mixing between materials [11] bring difficulties in extracting discriminative information from HSI for classification.

To address these issues, many feature extraction methods have been employed into hyperspectral classification in the past few decades [12]. In the early stage, many machine learning and handcrafted methods have been utilized to abstract the spatial-spectral information of HSI, such as gray-level co-occurrence matrix [13], morphological profiles [14], support vector machine (SVM) [15], and Gabor filters [16]. However, handcrafted features like morphological profiles heavily rely on professional expertise and are quite empirical, which may have poor adaptability to different classification scenes. In addition, most handcrafted methods only extract shallow features, which lead to unsatisfactory performances for complex image environments [17].

Encouraged by the successful application of deep learning in many fields, various deep learning methods, such as Stacked Autoencoder (SAE) [18], Deep Belief Network (DBN) [19], Recurrent Neural Network (RNN) [20] and Convolutional Neural Network (CNN) [21] have been applied into HSI classification. Compared to handcrafted methods, deep learning methods can extract deep features from the HSI, which shows a stronger ability to represent the geometrical attributes, spectral, and textural of the HSI data. Chen et al. [19] adopted unsupervised deep SAE to learn

Y. Ding et al., *Graph Neural Network for Feature Extraction and Classification of Hyperspectral Remote Sensing Images*, Intelligent Perception and Information Processing, https://doi.org/10.1007/978-981-97-8009-9_4

spatial and spectral features separately, which was the first time that the deep learning concept was introduced into the HSI classification task. Subsequently, the DBN and restricted Boltzmann machine were utilized for Hyperspectral images classification in [22]. Apart from the aforementioned unsupervised deep learning methods, some supervised deep learning models have been exploited for HSI classification. In [23], the recurrent neural network (RNN) was adopted to extract the contextual information among adjacent bands in HSIs. Then 1DCNN [24], 2DCNN [25], and 3DCNN [26] have also been proposed one after another. Recently, CNN-based methods are still an important research direction to solve HSIs classification tasks. Compared with the unsupervised methods, CNN-based HSI classification methods are able to extract efficient deep features from raw images directly, and yield better performance. However, a huge number of training samples is required in supervised methods, which cannot be satisfied for HSIs classification due to labeled training samples deficiency in HSIs. In addition, the CNN-based methods almost are applied to Euclidean data and cannot deal with non-Euclidean data [27, 28].

When CNN encountered a bottleneck in hyperspectral classification, the GCN method appeared. Recently, GCN is the hot spot in HSI classification [29–32]. In [33], Qin et al. have shown that the GCN-based semi-supervised method was a good solution to alleviate the problem of limitation labeled samples in HSI. The graph attention mechanism was introduced the algorithm by Sha et al. [34] to modify GCN. To extract the HSI spatial-spectral features, multiscale and context-aware GCN were introduced in [35, 36], respectively. To reduce the computational cost of GCN, Hong et al. [37] proposed a minibatch GCN and aimed to utilize the advantages of CNN and GCN comprehensively. However, there are still many defects in GCNs as following aspects:

- Firstly, GCNs are computational expensive (resulting from the graph Laplacian and the node aggregation mechanism), which is a significant bottleneck for HSI classification tasks, particularly processing large-scale HSI data sets.
- In addition, the traditional GCNs face the problem of over-smoothing (the representation of each node tends to be congruent) as the number of convolution layers becomes large (the number of convolution layers is generally less than three). For this reason, GCNs cannot be designed too deeply, which constrains the networks from extracting high-level features of the graph.
- Furthermore, to explore wider receptive fields and deeper features, more and more GCN layers have been stacked, which may bring a greater amount of calculation. Whether there is a way to expand the receptive field of graph neural network (GNN) while the number of aggregate nodes is reduced to reduce calculational complexity.
- Finally, the graph Laplacian matrix is adopted to define the importance of neighbors for each node in GCNs, which assign the same weight to each hop of nodes for the target node and cannot adjust the aggregate path automatically.

In response to the aforementioned problems, we proposed a novel multi-adaptive receptive field graph neural framework. In this paper, we introduce an adaptive path aggregate mechanism to explore a wider receptive field, which can learn the importance weight of different hop neighborhoods and aggregate the nodes selectively and automatically. And inspired by multi-scale in a visual field, we proposed multi-receptive field notion, which can extract multi-scale features in the graph. In summary, the main contributions of the paper can be expressed as threefold:

- We introduce an adaptive path aggregate mechanism to explore the receptive field of graph and suppress the impact of noise nodes on classification.
- A long short-term memory (LSTM) method is adopted to update the node and preserve the local convolutional feature of the node, which can alleviate the over-smoothing problem of graphs.
- We propose a novel multi-receptive field structure. The structure can extract multi-scale features in the graph.

4.2 Related Work and Preliminaries

In this part, some definitions and notations are introduced, which are related to the proposed method. We assume a graph $\mathcal{G} = (\mathcal{V}, \mathcal{E}, \boldsymbol{A})$. $\mathcal{E}$ presents the edge set, $\mathcal{V}$ denotes the vertex set, the adjacency matrix of the graph is $\boldsymbol{A} \in \mathbb{R}^{N\times N}$, and the weight of the edge between vertex i and vertex j is A_{ij}.

4.2.1 Graph Convolutional Networks

Kift and Welling [38] proposed GCN, which is the most representative graph convolution models. The Laplacian matrix $\boldsymbol{L}$ for a graph $\mathcal{G} = (\mathcal{V}, \mathcal{E}, \boldsymbol{A})$ is expressed as

$$\boldsymbol{L} = \boldsymbol{D} - \boldsymbol{A}, \tag{4.1}$$

where $\boldsymbol{D}$ is the degree matrix of the graph.

The corresponding symmetrically normalized Laplacian $\boldsymbol{L}_{\text{sym}}$ is as follows

$$\boldsymbol{L}_{sym} = \boldsymbol{I}_N - \boldsymbol{D}^{-\frac{1}{2}}\boldsymbol{A}\boldsymbol{D}^{-\frac{1}{2}}, \tag{4.2}$$

where $\boldsymbol{I}_N$ is an identity matrix.

Given the Fourier transform, the spectral decomposition $\boldsymbol{L}_{\text{sym}}$ can be written as

$$\begin{aligned}\boldsymbol{L}_{sym} &= \boldsymbol{U}\boldsymbol{diag}[\lambda_1, \ldots, \lambda_n]\boldsymbol{U}^T \\ &= \sum_{n=1}^{N} \lambda_n \boldsymbol{u}_n \boldsymbol{u}_n^T,\end{aligned} \tag{4.3}$$

where $\boldsymbol{U}$ is the eigenvectors matrix of $\boldsymbol{L}_{\text{sym}}$, i.e., $\boldsymbol{U}\boldsymbol{U}^T = E$; λ_n denotes the eigenvalue; and $\boldsymbol{u}_n$ represents the eigenvector.

Given a filter $g_{\boldsymbol{\theta}} = \text{diag}(\boldsymbol{\theta})$ and a graph signal $\boldsymbol{s} \in \mathbb{R}^N$ (graph node feature vector), We can perform the spectral convolution of $g_{\boldsymbol{\theta}}$ and $\boldsymbol{s}$ by the multiplication of a signal $\boldsymbol{s}$ with $\boldsymbol{g}_\theta$ the Fourier domain [39–41], namely

$$g_\theta \star \boldsymbol{s} = \boldsymbol{U} g_\theta(\boldsymbol{\Lambda}) \boldsymbol{U}^T \boldsymbol{s}, \tag{4.4}$$

where $\boldsymbol{\Lambda}$ and g_θ can be understood as the diagonal matrix of eigen-values of $\boldsymbol{L}_{sym}$ and an eigen-values function of $\boldsymbol{\Lambda}$, i.e., $\boldsymbol{g}_\theta(\boldsymbol{\Lambda})$, respectively. However, Eq. (4.4) requires massive calculation. The spectral convolution is simplified by setting the largest eigenvalue $\lambda_{max} = 2$ and approximating spectral filters with the 1th-order Chebyshev polynomials. The Eq. (4.4) can be rewritten as $g_\theta \star \boldsymbol{s} \approx \boldsymbol{\theta}(\boldsymbol{I}_N + \boldsymbol{L})\boldsymbol{x} = \boldsymbol{\theta}\left(\tilde{\boldsymbol{D}}^{\frac{1}{2}} \tilde{\boldsymbol{A}} \tilde{\boldsymbol{D}}^{-\frac{1}{2}}\right)\boldsymbol{x}$, $\tilde{\boldsymbol{D}}$ is the degree matrix of nodes, $\boldsymbol{I}_N + \boldsymbol{L}$ denotes an adjacency matrix. Formally, the spectral convolution defined by GCN over a graph is shown as follows

$$\boldsymbol{H}^{(l+1)} = \xi\left(\tilde{\boldsymbol{D}}^{\frac{1}{2}} \tilde{\boldsymbol{A}} \tilde{\boldsymbol{D}}^{-\frac{1}{2}} \boldsymbol{H}^{(l)} \boldsymbol{W}^{(l)}\right). \tag{4.5}$$

where ξ denotes an activation function, $\boldsymbol{W}^{(l)}$ is a trainable parameter matrix in layer $l+1$, $\boldsymbol{H}^{(l)}$ denotes the output of layer l and input of layer $l+1$.

4.2.2 Receptive Field of GCN

For a graph $\mathcal{G} = (\mathcal{V}, \mathcal{E}, \boldsymbol{A})$, the GCN calculations at the l-th layer graph network can be considered as $\boldsymbol{H}^{(l+1)} = \xi\left(\phi(A) H^{(l)} W^{(l)}\right)$ as Eq. (4.5), which defines embeddings of graph nodes at the l-th layer. To simplify the analysis, we can ignore the activation function and treat $H^{(l)}$ and $W^{(l)}$ as constants, Eq. (4.5) can be rewritten as

$$\boldsymbol{H}^{(1)} = \phi(A) H^{(0)} W^{(1)}, \tag{4.6}$$

Aftering l-layer convolution, Eq. (4.6) can be formed as

$$\boldsymbol{H}^{(l)} = \phi(A)^l H^{(0)} W^{(1)}. \tag{4.7}$$

where $\phi(A)^l$ is the l-th order pre-defined receptive field [42]. Equation (4.7) demonstrates that the extent of neighbors to exploit are determined by the depth of layers l.And if node j satisfies $d(i, j) \leq l$ (shortest path distance between node i and j), it will be embedded by node i by pre-defining the importance weight $\phi(A)_{ij}^{l}$. Fundamentally, the target node receptive field in a graph domain is equivalent to the subgraph that consists of nodes along paths to the target node.

As the analysis above, we find that GCN pre-defines the same weight to the nodes in each hop layer. The method using distance to determine the aggregation weight has great defects in practical applications, because the nodes may be noisy. In reality, the correlations between different nodes and the target node are not the same. To ameliorate these issues, we should design an effective aggregator function, which can transmit signals around the l-hop neighborhood of each node.

4.3 Proposed Method

In this part, first, the overview of our MARP is presented. Then we explain the adaptive path aggregate mechanism for the graph. Afterward, multi-receptive field manipulation is illustrated. Finally, we introduce the proposed HSI classification framework and our training strategy.

4.3.1 Overview of the Proposed Method

In this section, the overview of MARP will be illustrated (see Fig. 4.1). Firstly, the input HSI will be segmented into superpixels by adopting unsupervised principal component analysis (PCA) [43] and simple linear iterative clustering (SLIC) [43] method, which can reduce the numbles of graph nodes for subsequent graph construction. Simultaneously, spectral features of each superpixel are calculated by employing mean operation. Then, the graph is constructed. In the following, we propose a multi-adaptive receptive field method to extract graph features. Different receptive field paths can extract different graph features. Finally, the multi-scale features are used for node classification by employing cross-entropy loss.

4.3.2 Adaptive Receptive Path Aggregation (ARP)

In practice, we find that graphs may contain noise nodes and different nodes make different contributions to the target node. Thus, the most suitable receptive field should be found to present the target node, instead of treating all neighbor nodes equally in GCNs. The motivation of the adaptive receptive field path is shown in Fig. 4.2.

Inspired by the graph attention mechanism, we proposed a node attention method to define the importance of any 1-hop neighbors and learn the 1-hop neighbor (hidden) features. The relationship between any two nodes can be expressed as

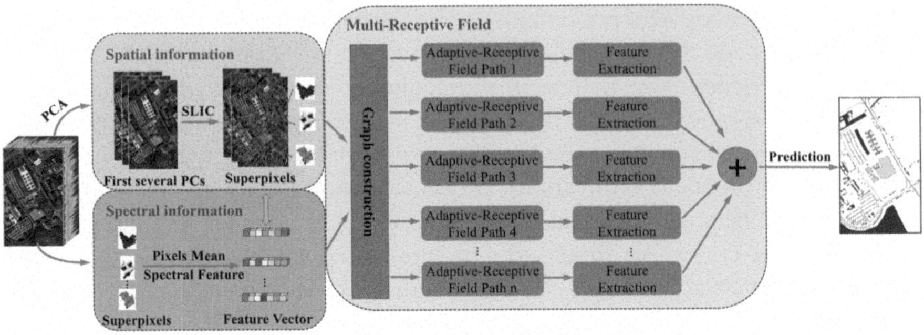

Fig. 4.1 The flowchart of the proposed multi-adaptive receptive field network. The MARP can be divided into three parts: graph construction, feature extraction, and prediction. The different color boxes in the figure present the different processes of the algorithm. We treat superpixels as graph nodes as constructing the graph, the feature vector of the graph node is calculated by the average of the involved pixels spectral feature in the superpixel. '+' denotes concatenate operation

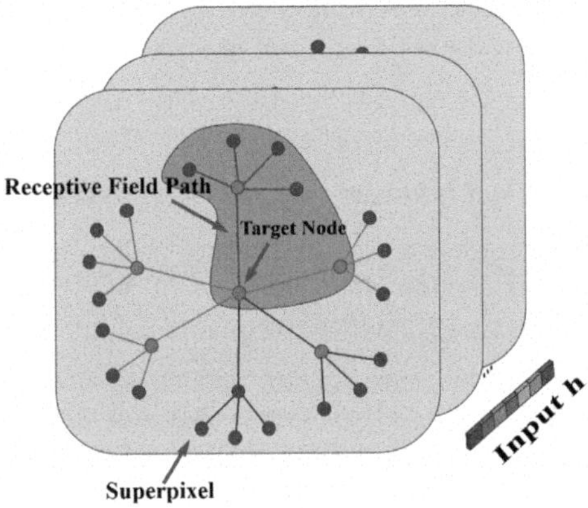

Fig. 4.2 The motivation of adaptive receptive field path. In the figure, the superpixels are treated as the graph nodes. Different colors of the line between two nodes denote different relationships

$$e_{ij} = \left(LeakyReLU\left(v^T\left[\boldsymbol{Wh}_i||\boldsymbol{Wh}_j\right]\right)\right) \tag{4.8}$$

The importance weight of node j to node i is shown as Eq. (4.8), $LeakyReLU(\cdot)$ is an activation layer, $||$ denotes concatenation operation, and v^T is a trainable parameter vector of the network.

A softmax function is adopted for Eq. 4.8 normalization

$$a\left(\boldsymbol{h}_i, \boldsymbol{h}_j\right) = \frac{\exp\left(LeakyReLU\left(v^T\left[\boldsymbol{Wh}_i||\boldsymbol{Wh}_j\right]\right)\right)}{\sum_{k \in N_i} \exp\left(LeakyReLU\left(v^T\left[\boldsymbol{Wh}_i||\boldsymbol{Wh}_j\right]\right)\right)} \tag{4.9}$$

We aim to realize the aggregation of neighbor nodes by learning the importance of 1-hop neighbor. The graph node can be adaptive update as

$$h_i^{(tmp)} = LeakyReLU\left(W^{(t)^T} \sum_{j \in \mathcal{N}(i) \cup \{i\}} \alpha\left(h_i^{(t)}, h_j^{(t)}\right) \cdot h_j^{(t)}\right) \tag{4.10}$$

where $\mathcal{N}(i)$ is 1-hop neighborhood node of node i, $h_i^{(t)}$ denotes the node feature vector of node i, $W^{(t)^T}$ is a trainable parament, which aims to perform a linear transformation for the feature of aggregate neighbors to make them more powerful. Equation (4.10) shows the aggregation principle of node i with 1-hop nodes. $h_i^{(tmp)}$ contains the relationships between all neighbor nodes of node i and can suppress the influence of noisy nodes on the aggregation process.

Both shallow features and deep features all have an important impact on classification [44]. In reality, the node features in the intermediate process can be regarded as a set of timing signals. Inspired by timing signal processing, an LSTM (shown as Fig. 4.3) is proposed. The LSTM mathematical expression of the processing process in our proposed method is as follows

$$New\ aggregate\ signal : \tilde{C} = \tanh\left(W_c^{(t)^T} h_i^{(tmp)}\right), \tag{4.11}$$

$$Updategate : i_i = \sigma\left(W_i^{(t)^T} h_i^{(tmp)}\right), \tag{4.12}$$

$$Forgetgate : f_i = \sigma\left(W_f^{(t)^T} h_i^{(tmp)}\right), \tag{4.13}$$

$$Memoryupdate : C_i^{(t+1)} = f_i \odot C_i^t + i_i \odot \tilde{C}, \tag{4.14}$$

$$O_i = \sigma\left(W_o^{(t)^T} h_i^{(tmp)}\right), \tag{4.15}$$

$$Output : h_i^{(t+1)} = O_i \odot \tan h\left(C_i^{(t+1)}\right). \tag{4.16}$$

where $W_c^{(t)^T}$, $W_i^{(t)^T}$, $W_f^{(t)^T}$, $W_o^{(t)^T}$ are trainable paraments in LSTM network, $\tan h$ is a nonlinear layer, $\odot$ denotes multiplication operation, $\sigma \in (0,1)$ is forgotten coefficient, and $C_i^t \in (0,1)$ is memory coefficient, which can preserve the previous features.

Algorithm1: Adaptive receptive path aggregation.

Input: Input graph; aggregate depth L ; node feature h_i^0.

Output: $h_i^{(t+1)}$.

```
1:  for   l = 1 to L do
2:          Define the importance of t-hop neighbors by Eq. (4.9)
3:          Aggregate the t-hop neighbors by Eq. (4.10)
4:          Memory update and preserve the t layer features by Eq. (4.17)
5:          Graph network node update and t layer output h_i^(t+1)
6:          Loss ℒ backpropagation
7:  end
8:  return h_i^(t+1).
```

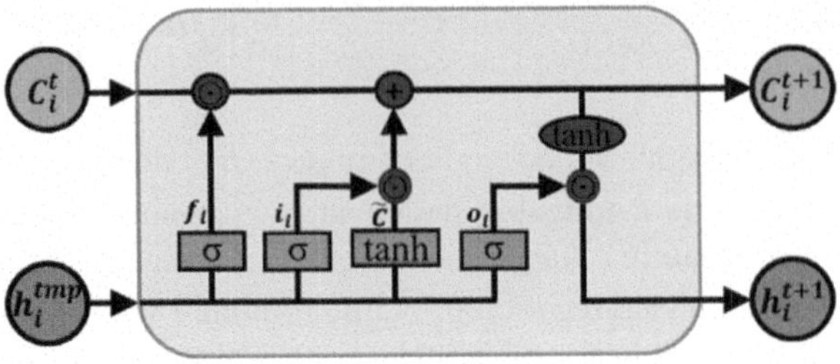

Fig. 4.3 The overview framework of LSTM

As described above, we can control the aggregated hops by adjusting the number of adaptive convolutional layers. To extract the neighbor node features in the T order, i.e., $\{h_i^0, h_i^1, h_i^2, \cdots h_i^T\}$. The aggregate output of each hop will be input to LSTM for processing, just like time-series signals. With this projection, we have

$$\begin{aligned} \boldsymbol{h}_i^{(t+1)} = & \left(\mathbf{W}_o^{(t)^T} \boldsymbol{h}_i^{(tmp)}\right) \odot \tan h\Big(\sigma\left(\mathbf{W}_f^{(t)^T} \boldsymbol{h}_i^{(tmp)}\right) \odot \boldsymbol{C}_i^t \\ & + \sigma\left(\mathbf{W}_i^{(t)^T} \boldsymbol{h}_i^{(tmp)}\right) \odot \tan h\left(\mathbf{W}_c^{(t)^T} \boldsymbol{h}_i^{(tmp)}\right)\Big) \end{aligned} \tag{4.17}$$

Therefore, the aggregated features of each hop will be saved for node prediction. The whole process details of adaptive receptive path aggregation are shown in Algorithm 1.

4.3.3 *Multi-receptive Field Manipulation*

We describe the adaptive receptive path aggregate mechanism in Sect. 4.3.2, in this part we will present the motivation of adopting multi-receptive field and implementation details. As shown in Algorithm 1, we find that node feature h_i^0 and 1-hop nodes play a significant role in process of aggregation, which determines whether the algorithm can find the global optimal receptive field and extract global features for HSI classification. Specifically, the algorithm may fall into a local optimum. Multiscale has been proven to be able to effectively extract the global information of the image [45, 46]. In our method, we propose a multi-branch adaptive receptive path aggregation layer to capture multi-scale spectral-spatial information of HSI. Specifically, the branch n of node i is connected to its n-hop neighbors at first, which can avoid the negative impact of low-order hop nodes and learns long-distance feature information of HSI. The multi-receptive field design is exhibited in Fig. 4.4. By considering both the efficiency and effectiveness, in the proposed method, the graph is constructed at 1, 2, and 3 hops. Based on Eq. (4.14), then the out the output of MARP can be expressed as

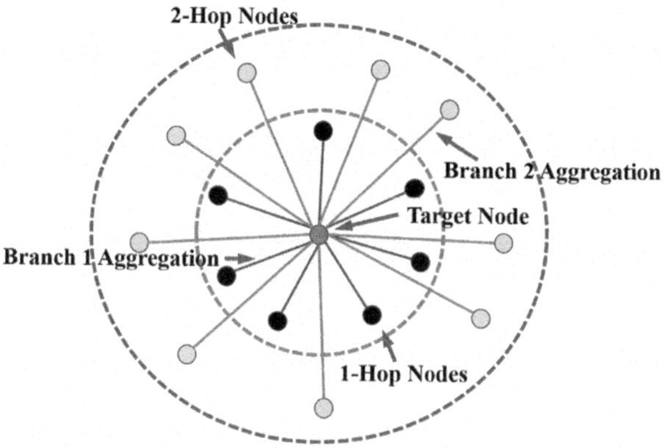

Fig. 4.4 The illustration of multi-receptive field design. The figure shows the 1-hop nodes and 2 hop nodes of h_i.The 1-hop neighbors of h_i is demonstrated as black, 2-hop neighbors of h_i are presented as yellow. Different colors of lines denote the aggregation of jump points on different layers

$$\boldsymbol{O}(\boldsymbol{h}_i) = \boldsymbol{H}_1(\boldsymbol{h}_i) \cup \boldsymbol{H}_2(\boldsymbol{h}_i) \cup \boldsymbol{H}_3(\boldsymbol{h}_i) \cdots \cup \boldsymbol{H}_n(\boldsymbol{h}_i). \tag{4.18}$$

where $\boldsymbol{H}_1(\boldsymbol{h}_i)$, $\boldsymbol{H}_2(\boldsymbol{h}_i)$, $\boldsymbol{H}_3(\boldsymbol{h}_i)$, and $\boldsymbol{H}_n(\boldsymbol{h}_i)$ are the output of branch 1, branch 2, branch 3, and branch n, respectively. And $\cup$ denotes the concatenate operation.

4.3.4 HSI Classification Using MARP

(1) *HSI Preprocessing.* Huge number of pixels are contained by HSI in the spatial dimension, which brings a lot of computing burden to subsequent network processing. To address this problem, image segmentation technology is adopted to divide the HSI into small blocks, and the pixels contained in the small blocks have a strong spatial-spectral correlation. Assume an HSI dataset $\{x_1, x_2, \cdots, x_m\}$ with B bands and m pixels. The superpixel map can be mathematically expressed as

$$HSI = \cup_{i=1}^{\mathcal{N}} S_i,\ S_i \cap S_j = \varnothing,\ i \neq j;\ i, j = 1,2, \ldots, \mathcal{N}. \tag{4.19}$$

where $\mathcal{N}$ is the total number of superpixels, S_i denotes a superpixel containing n_i pixels.

Because the classical segmentation method was originally designed for segmenting RGB images, it is impossible to segment HSI into superpixels by using the classical segmentation algorithms directly. Therefore, to split the HSI into superpixels, it is necessary to perform a spectral dimensionality reduction on the HSI in advance. In our proposed method, we employ a popular unsupervised principal component analysis (PCA) [43] method to reduce the spectral dimension of the HSI, and the first principal component (contains abundant original HSI information) is taken to generate the base image. Then we employ the simple linear iterative clustering (SLIC) [43] to segment the dimension reduced false-color image. Finally, the feature vector of the graph node is calculated by the average of the involved pixels spectral feature in the superpixel, which can suppress the influence of noise. The overview of the HSI preprocessing is shown as Fig. 4.1. The input feature vector of superpixel $\boldsymbol{h}_i$ can be rewritten as

$$\boldsymbol{h}_i = \frac{1}{n_i}\sum\nolimits_{k \in n_i} \boldsymbol{h}_k^p, \tag{4.20}$$

where n_i and $\boldsymbol{h}_k^p$ are the number and the spectral feature of pixel contained in superpixel S_i,respectively.

(2) *Graph Construction.* The proposed MARP method generates node embeddings by gradually fusing the features in the graph. In this paper, we treat superpixels as graph nodes to improve computational efficiency. For a graph $\mathcal{G} = (\mathcal{V}, \mathcal{E}, \boldsymbol{A})$. the adjacency matrix $\boldsymbol{A}_{ij} \in \mathbb{R}^{N \times N}$ can be expressed as

$$a(\boldsymbol{h}_i, \boldsymbol{h}_j) = \frac{\exp\left(LeakyReLU\left(\boldsymbol{v}^T\left[\boldsymbol{W}\boldsymbol{h}_i || \boldsymbol{W}\boldsymbol{h}_j\right]\right)\right)}{\sum_{k \in N_i} \exp\left(LeakyReLU\left(\boldsymbol{v}^T\left[\boldsymbol{W}\boldsymbol{h}_i || \boldsymbol{W}\boldsymbol{h}_j\right]\right)\right)},$$
$$\boldsymbol{h}_i \in \mathcal{N}_t(\boldsymbol{h}_j) \text{ or } \boldsymbol{h}_j \in \mathcal{N}_t(\boldsymbol{h}_i). \tag{4.21}$$

where $\boldsymbol{h}_i$ and $\boldsymbol{h}_j$ denote the spectral feature of nodes i, j (i.e., superpixels in the method), $\mathcal{N}_t(\boldsymbol{h}_j)$ denotes the t-hop neighbors set of $\boldsymbol{h}_i$.

Algorithm 2: MARP for HSI Classification.

Input: Input HSI; number aggregate branches $\boldsymbol{b}$; number of convolutional layers $\boldsymbol{L}$ of each branch; number of epochs $\boldsymbol{T}$.

Output: Pixel label.

The whole image is segmented into superpixel via PCA-SLIC algorithm;
The superpixels input features are extracted and the graph $\boldsymbol{A}$ is constructed by Eq. (4.20) and Eq. (4.21);
for t=1 to T do
  for l = 1 to L do
    The outputs of the lth layer output $h^{(lmp)}$ calculated by Eq. (4.10);
    The graph nodes $h^{(l)}$ are updated by Eq. (4.17);
  end
  The MARP output $\boldsymbol{O}(\boldsymbol{h}_i)$ is calculated by Eq. (4.18);
  The error term is calculated by Eq. (4.22) and the weight matrices are updated by Adam gradient descent;
end
Perform label prediction for each pixel.
Output: each pixel label.

MARP Implementation for HSI Classification. Equation (4.21) shows the graph construction mechanism, which is the base graph for subsequence network extracting. Then a multi-adaptive receptive field network is proposed to extract global features of the graph adaptively. In our MARP, we employ cross-entropy error to penalize the difference between the MARP output and the label examples, namely,

$$L = -\sum_{z \in \mathbf{y}_\mathrm{G}} \sum_{f=1}^{C} \boldsymbol{Y}_{zf} ln \boldsymbol{O}_{Gzf}^{(final)}. \tag{4.22}$$

where $\boldsymbol{Y}_{zf}$ denotes the label matrix, $\mathbf{y}_\mathrm{G}$ is the labeled example set, and C presents the number of landcover categories, $\boldsymbol{O}_{Gzf}^{(final)}$ is the output of MARP. Furthermore, we adopt Adam [47] gradient descent to update the weight of MARP. The implementation process of MARP for HSI classification is demonstrated in Algorithm 2.

4.4 Experiments

To assess the superior of MARP for HSI classification tasks, extensive experimental comparison and analysis are shown in this paper. Firstly, three HSI datasets used in our experiments are introduced. Then, we compared the proposed MARP with six state-of-the-art methods on three real HSI data sets. Furthermore, the impact of hyperparameters on MARP performance is analyzed. Subsequently, the ablative effects of multi-receptive field and ARP manipulation have been studied. Finally, the comparative running times are demonstrated.

4.4.1 Dataset Processing and Description

We introduce four widely used datasets, including Pavia University, Salinas, and Houston 2013 in this part. 30 labeled pixels from each class are randomly selected to train the model, and the remaining pixels are adopted for testing.

(1) *Pavia University:* The Pavia University was acquired by an airborne reflection optics spectral imaging system (ROSIS) in 2003, which is part of the Pavia City of Italy HSI data. The dataset contains 103 bands after eliminating 12 noisy bands. The size of the data set is 610×340. The whole map can be categorized into nine kinds of features, the false color image and ground truth map are shown in Fig. 4.5. The details of each class are presented in Table 4.1.

(2) *Salinas:* The Salinas was captured by AVIRIS sensors in Salinas Valley, California. The spatial resolution of the data is 3.7 m and the size is 512×217. There are 224 bands in the original data, and 204 bands are left after removing the water vapor absorption bands. The data contains 16 crop categories. The false color image and ground truth map are shown in Fig. 4.6. The details of each class are presented in Table 4.2.

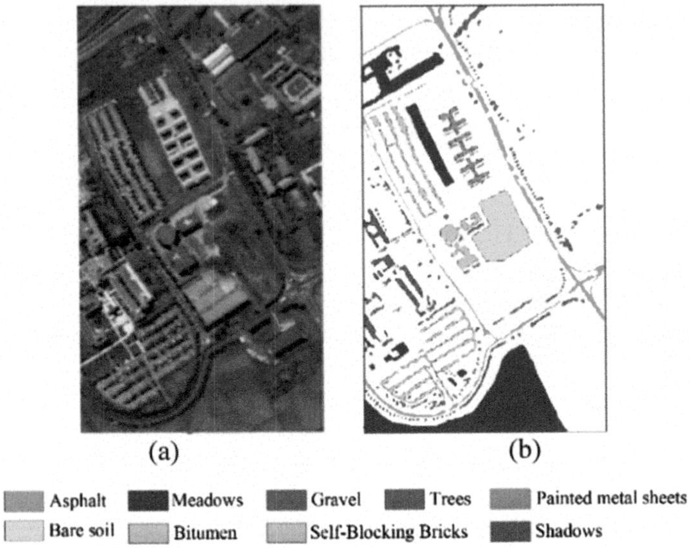

Fig. 4.5 Pavia University. **a** False color image. **b** Ground truth map

Table 4.1 Amounts of training and testing pixels in Pavia University dataset

No.	Class name	Training	Testing
1	Asphalt	30	6601
2	Meadows	30	18,619
3	Gravel	30	2069
4	Trees	30	3034
5	Painted-metal-sheets	30	1315
6	Bare-soil	30	4999
7	Bitumen	30	1300
8	Self-blocking-bricks	30	3652
9	Shadows	30	917

(3) *Houston 2013:* The Houston 2013 data set was obtained by ITRES CASI-1500 sensor and provided by 2013 IEEE GRSS data fusion competition. The data size is 349 × 1905, including 144 bands ranging from 364 to 1046 nm. Ground cover is labeled as 15 categories, which are shown in Fig. 4.7. The details of each class are presented in Table 4.3.

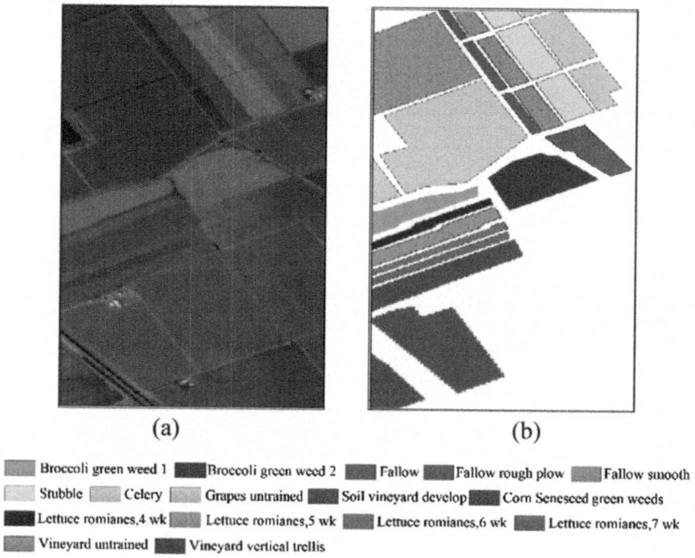

Fig. 4.6 Salinas. **a** False color image. **b** Ground truth map

Table 4.2 Amounts of training and testing pixels in Salinas dataset

No.	Class name	Training	Testing
1	Broccoli-green-weed 1	30	1979
2	Broccoli-green-weed 2	30	3696
3	Fallow	30	1946
4	Fallow-rough-plow	30	1364
5	Fallow-smooth	30	2648
6	Stubble	30	3929
7	Celery	30	3549
8	Grapes-untrained	30	11,241
9	Soil-vineyard-develop	30	6137
10	Corn-Senesced-green-weeds	30	3248
11	Lettuce-romianes-4 week	30	1038
12	Lettuce-romianes-5 week	30	1897
13	Lettuce-romianes-6 week	30	886
14	Lettuce-romianes-7 week	30	1040
15	Vineyard-untrained	30	7238
16	Vineyard-vertical-trellis	30	1777

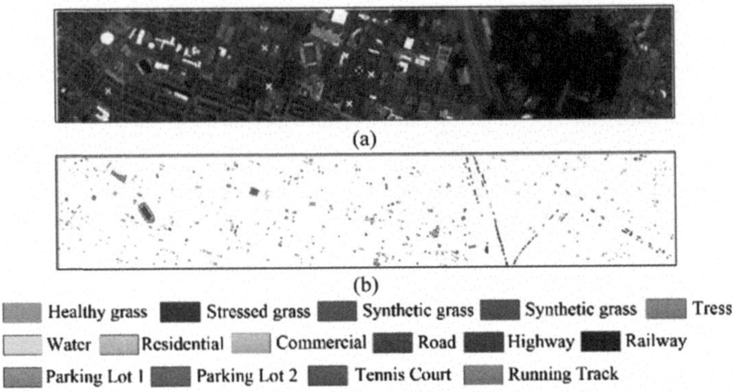

Fig. 4.7 Houston 2013. **a** False color image. **b** Ground truth map

Table 4.3 Amounts of training and testing pixels in Houston 2013 dateset

No.	Class name	Training	Testing
1	Healthy-grass	30	1221
2	Stressed-grass	30	1224
3	Synthetic-grass	30	667
4	Tress	30	1214
5	Soil	30	1212
6	Water	30	295
7	Residential	30	1238
8	Commercial	30	1214
9	Road	30	1222
10	Highway	30	1197
11	Railway	30	1205
12	Parking-Lot-1	30	1203
13	Parking-Lot-2	30	439
14	Tennis-court	30	398
15	Running-track	30	630

4.4.2 Setup

In this paper, all experiment is conducted via PyTorch. The hyperparameters selection in our MARP, including segmentation scale N, learning rate lr, number of MARP layers L, the multi-scale number of MARP S and number of iterations T, are demonstrated in Table 4.4.

(1) *Evaluation Index:* To evaluate the performance of the MARP model, four widely accepted indexes are used, i.e., Overall accuracy (OA), Per-class accuracy (PA), Average accuracy (AA), and Kappa coefficient (Kappa).

Table 4.4 Hyperparameter settings for different datasets

Dataset	N	lr	L	S	T
Pavia University	10,000	0.001	4	3	500
Salinas	1000	0.001	4	3	500
Houston 2013	20,000	0.001	4	3	500

Table 4.5 Architectural details of proposed network

Module	Details		
Pixel-to-region assignment	PCA SLIC	Spectral information of each region (input)	
Graph construction	Calculate adjacency matrix $\boldsymbol{A}$ of graph		
Original graph processing	MARP-1(4) Branch1, layers = 4	MARP-2(4) Branch2, layers = 4	MARP-3(4) Branch2, layers = 4
Feature fusion	Concatenation BN		
Output	Cross-entropy (target categories)		

(2) *Compared Methods:* To verify the classification accuracy and real-time performance of the proposed MARP, two CNN-based methods, i.e., convolutional RNN (CRNN) [21] and diverse region based deep CNN (DR-CNN) [20], two GNN methods, i.e., graph sample and aggregate-attention (SAGE-A) [8] and multiscale dynamic GCN (MDGCN) [35], and two traditional machine learning methods, i.e., joint collaborative representation and SVM with decision fusion (JSDF) [15] and multiband compact texture units (MBCTU) [14] are employed. The architectural details of the MARP are shown in Table 4.5. In our experiment, all the aforementioned methods are run 10 times, and the performance of the methods is measured by the above four indexes. The experiments are carried out with a 3.70G Intel i9-10900K CPU and a GeForce GTX 1080Ti 11G GPU.

4.4.3 Comparison Results

(1) *Quantitative Results*

In this section, the classification performance of MARP is quantitatively evaluated by comparing the proposed MARP with the aforementioned six algorithms. The quantitative results are reported in Tables 4.6, 4.7, 4.8, and 4.9. From the results, we conclude that our MARP outperforms the compared algorithms.

In Table 4.6, GNN-based methods, i.e., SAGE-A, MDGCN, and MARP, demonstrate their superiority over CNN-based methods. This is because the GNN-based method can learn the relationship between nodes, which can extract more diverse classification features under the same training sample conditions. At the same time,

Table 4.6 Quantitative classification results of the Pavia University dataset with thirty labeled training pixels per class

Class No.	CNN-PPF	DR-CNN	CRNN	SAGE-A	MDGCN	JSDF	MBCUT	MARP
1	**97.66 ± 1.21**	92.10 ± 3.67	79.85 ± 4.62	96.50 ± 1.28	93.55 ± 0.37	82.40 ± 4.07	87.49 ± 3.99	95.59 ± 3.21
2	94.34 ± 2.47	96.39 ± 1.85	82.33 ± 3.17	99.17 ± 0.62	**99.25** ± 0.23	90.76 ± 3.74	89.11 ± 5.58	98.14 ± 1.42
3	92.37 ± 3.22	84.23 ± 4.21	89.67 ± 3.69	**98.26 ± 1.77**	92.03 ± 0.24	86.71 ± 4.14	86.24 ± 4.23	96.40 ± 2.13
4	76.64 ± 5.21	95.26 ± 1.43	91.45 ± 2.44	92.85 ± 2.92	83.78 ± 1.55	92.88 ± 2.16	90.61 ± 3.39	**95.75 ± 1.49**
5	67.14 ± 4.19	97.77 ± 1.66	94.12 ± 1.78	97.49 ± 2.43	99.47 ± 0.09	**100.00 ± 0.00**	97.18 ± 1.18	**100.00 ± 0.00**
6	98.21 ± 0.23	90.44 ± 4.28	91.37 ± 2.11	94.78 ± 1.62	95.26 ± 0.50	94.30 ± 4.55	93.25 ± 2.93	**97.62 ± 1.78**
7	98.83 ± 0.09	89.05 ± 2.85	93.67 ± 1.97	97.92 ± 1.48	98.92 ± 1.04	96.62 ± 1.37	93.49 ± 2.47	**99.49 ± 0.58**
8	96.55 ± 1.07	78.49 ± 4.71	80.18 ± 3.29	89.84 ± 4.29	94.99 ± 1.33	94.69 ± 3.74	84.14 ± 4.78	**95.21 ± 2.42**
9	96.12 ± 2.42	96.34 ± 1.09	82.34 ± 4.86	98.80 ± 0.82	81.03 ± 0.49	**99.56 ± 0.36**	96.57 ± 1.22	98.03 ± 1.77
OA (%)	89.35 ± 1.35	92.62 ± 1.83	85.46 ± 1.75	96.19 ± 1.21	95.68 ± 0.22	90.82 ± 1.30	89.43 ± 2.14	**97.18 ± 0.92**
AA (%)	90.87 ± 1.39	91.12 ± 1.58	87.22 ± 1.82	96.18 ± 0.92	93.15 ± 0.28	93.10 ± 0.65	90.90 ± 0.89	**97.36 ± 1.17**
Kappa	86.01 ± 1.71	90.26 ± 1.72	84.19 ± 1.43	96.24 ± 0.98	94.25 ± 0.29	88.02 ± 1.62	86.24 ± 2.62	**97.69 ± 0.83**

Best performances are shown in bold

Table 4.7 Quantitative classification results of the Salinas dataset with thirty labeled training pixels per class

Class No.	CNN-PPF	DR-CNN	CRNN	SAGE-A	MDGCN	JSDF	MBCUT	MARP
1	98.80 ± 0.12	99.40 ± 0.42	99.34 ± 0.53	99.78 ± 0.21	99.98 ± 0.03	**100.00 ± 0.00**	99.18 ± 0.80	**100.00 ± 0.00**
2	**100.00 ± 00**	99.46 ± 0.39	99.17 ± 0.21	100.00 ± 0.00	99.90 ± 0.28	**100.00 ± 0.00**	99.76 ± 0.33	**100.00 ± 0.00**
3	91.53 ± 2.13	98.58 ± 1.42	96.54 ± 1.39	99.45 ± 0.35	99.80 ± 0.21	**100.00 ± 0.00**	99.13 ± 1.04	**100.00 ± 0.00**
4	**100.00 ± 00**	99.70 ± 0.17	97.32 ± 0.27	100.00 ± 0.00	97.49 ± 2.16	99.93 ± 0.09	97.61 ± 0.82	**100.00 ± 0.00**
5	92.22 ± 1.47	98.90 ± 1.01	98.75 ± 0.89	98.75 ± 1.17	97.96 ± 0.77	**99.77 ± 0.31**	96.54 ± 1.01	99.21 ± 0.28
6	99.97 ± 0.01	99.57 ± 0.38	99.19 ± 0.77	89.72 ± 3.62	99.10 ± 1.67	**100.00 ± 0.00**	99.74 ± 0.32	99.74 ± 0.13
7	99.72 ± 0.11	99.50 ± 0.42	98.67 ± 1.30	**100.00 ± 0.00**	98.18 ± 1.49	99.99 ± 0.01	98.26 ± 1.64	98.62 ± 0.72
8	**100.00 ± 00**	75.59 ± 6.72	72.38 ± 3.98	85.25 ± 5.27	92.78 ± 4.61	87.79 ± 4.89	81.98 ± 4.32	94.16 ± 1.71
9	99.98 ± 0.01	99.75 ± 0.19	97.29 ± 0.61	95.31 ± 2.41	**100.00 ± 0.00**	99.67 ± 0.33	99.47 ± 0.51	99.82 ± 0.15
10	97.52 ± 1.69	94.29 ± 1.90	91.44 ± 1.64	97.18 ± 0.82	**98.31 ± 1.29**	96.53 ± 2.55	92.21 ± 2.75	94.68 ± 2.78
11	96.24 ± 2.11	97.57 ± 0.91	96.82 ± 0.78	96.36 ± 1.42	99.39 ± 0.55	**99.76 ± 0.21**	96.24 ± 2.68	98.27 ± 1.62
12	98.80 ± 0.57	99.99 ± 0.01	99.21 ± 0.32	99.18 ± 0.27	99.01 ± 0.78	**100.00 ± 0.00**	98.98 ± 0.45	98.92 ± 0.73
13	70.49 ± 6.33	99.95 ± 0.05	97.29 ± 0.86	97.26 ± 1.63	97.59 ± 1.32	**100.00 ± 0.00**	96.73 ± 1.66	97.66 ± 1.83
14	77.72 ± 5.91	98.57 ± 0.28	95.10 ± 1.73	**99.13 ± 0.46**	97.92 ± 1.72	98.71 ± 0.72	96.50 ± 3.05	92.13 ± 3.21
15	25.65 ± 6.29	72.18 ± 4.97	76.33 ± 4.62	87.23 ± 3.77	95.71 ± 4.57	81.86 ± 5.26	79.41 ± 5.67	**96.82 ± 0.43**
16	99.78 ± 0.19	98.45 ± 0.83	97.99 ± 0.61	98.47 ± 0.81	98.18 ± 2.92	98.99 ± 0.63	96.89 ± 2.19	**98.75 ± 0.79**
OA (%)	88.03 ± 1.26	90.35 ± 1.67	87.64 ± 0.83	92.82 ± 1.00	97.25 ± 0.87	94.67 ± 0.77	92.14 ± 0.86	**97.81 ± 0.88**
AA (%)	90.52 ± 1.41	95.72 ± 0.41	94.55 ± 0.57	96.12 ± 0.82	**98.21 ± 0.30**	97.69 ± 0.34	95.54 ± 0.56	98.04 ± 0.29
Kappa	86.55 ± 2.07	89.26 ± 1.30	86.72 ± 1.01	93.26 ± 0.76	96.94 ± 0.96	94.06 ± 0.85	91.25 ± 0.95	**97.12 ± 0.58**

Best performances are shown in bold

Table 4.8 Quantitative classification results of the Houston 2013 dataset with thirty labeled training pixels per class

Class No.	CNN-PPF	DR-CNN	CRNN	SAGE-A	MDGCN	JSDF	MBCUT	MARP
1	96.75 ± 1.17	95.62 ± 1.31	82.45 ± 2.19	86.17 ± 3.26	93.42 ± 4.25	97.41 ± 1.21	92.86 ± 3.83	**97.67 ± 1.01**
2	81.75 ± 4.62	96.78 ± 2.38	84.12 ± 3.64	94.42 ± 2.41	93.67 ± 3.60	**99.84 ± 0.25**	92.18 ± 2.79	92.51 ± 2.40
3	**100.00 ± 00**	96.75 ± 1.40	91.56 ± 1.07	98.80 ± 1.17	98.12 ± 1.09	99.88 ± 0.22	97.42 ± 1.19	98.81 ± 1.19
4	97.01 ± 1.67	93.41 ± 3.62	91.29 ± 3.78	96.19 ± 2.15	95.58 ± 1.85	**98.22 ± 2.80**	90.96 ± 1.98	92.46 ± 3.16
5	85.92 ± 6.21	99.15 ± 0.57	98.81 ± 0.62	**100.00 ± 0.00**	99.00 ± 1.30	**100.00 ± 0.00**	97.17 ± 1.29	99.47 ± 0.53
6	48.36 ± 5.96	93.83 ± 1.82	94.83 ± 2.19	93.72 ± 2.35	93.28 ± 6.08	**99.32 ± 1.09**	91.78 ± 3.22	96.82 ± 2.67
7	84.92 ± 3.59	80.71 ± 5.81	86.42 ± 3.42	**93.31 ± 3.72**	87.68 ± 4.41	91.32 ± 4.91	82.88 ± 3.81	88.25 ± 1.27
8	**96.24 ± 2.21**	78.32 ± 4.65	53.05 ± 8.71	84.63 ± 3.98	80.45 ± 6.12	68.82 ± 6.16	71.85 ± 5.64	86.44 ± 4.63
9	84.35 ± 3.79	76.90 ± 3.86	84.04 ± 4.63	**96.74 ± 2.36**	89.64 ± 2.26	69.47 ± 8.56	81.94 ± 4.25	90.17 ± 2.61
10	80.18 ± 3.78	81.99 ± 2.61	45.14 ± 8.77	**97.59 ± 1.24**	90.06 ± 6.41	85.63 ± 9.32	87.31 ± 5.08	89.71 ± 6.12
11	89.30 ± 5.98	84.04 ± 0.88	61.85 ± 9.39	89.39 ± 5.97	86.73 ± 3.22	94.51 ± 3.82	77.41 ± 6.46	**96.88 ± 3.18**
12	85.70 ± 7.35	81.92 ± 1.37	84.40 ± 2.93	**92.98 ± 1.62**	89.44 ± 5.69	84.33 ± 5.33	86.35 ± 5.85	84.82 ± 5.23
13	83.80 ± 4.27	86.54 ± 4.17	84.14 ± 3.12	85.69 ± 5.69	92.78 ± 4.45	**98.10 ± 1.28**	85.58 ± 5.35	93.27 ± 5.28
14	87.69 ± 5.23	99.31 ± 0.16	96.03 ± 0.73	**100.00 ± 0.00**	99.43 ± 0.97	**100.00 ± 0.00**	96.85 ± 1.85	**100.00 ± 0.00**
15	97.73 ± 1.22	99.60 ± 0.29	93.45 ± 2.41	98.46 ± 1.46	96.27 ± 1.72	**99.86 ± 0.36**	92.27 ± 3.32	96.53 ± 3.47
OA (%)	86.72 ± 1.03	88.08 ± 1.47	82.10 ± 1.21	92.76 ± 0.74	91.40 ± 0.92	90.51 ± 0.95	87.07 ± 1.12	**93.02 ± 0.62**
AA (%)	90.52 ± 0.97	89.66 ± 1.17	79.21 ± 1.02	**93.87 ± 1.01**	92.37 ± 0.89	92.46 ± 0.75	89.32 ± 1.08	93.58 ± 0.79
Kappa	86.55 ± 2.09	87.10 ± 0.99	77.61 ± 1.19	93.25 ± 0.92	90.70 ± 1.00	89.74 ± 1.03	86.01 ± 1.21	**93.41 ± 1.43**

Best performances are shown in bold

Table 4.9 The classification quality metrics obtained using GNN-based models with noisy data

SAGE-A

η_p	Gaussian				Impulsive				Poisson			
	0.1	0.2	0.3	0.4	0.1	0.2	0.3	0.4	0.1	0.2	0.3	0.4
OA (%)	95.57	94.82	94.53	94.18	95.02	93.77	92.56	90.82	95.16	93.29	91.72	90.66
AA (%)	95.51	94.63	94.50	94.39	94.82	93.81	92.70	90.77	94.82	93.11	91.62	89.26
Kappa	95.32	94.29	94.03	94.02	94.93	93.66	92.48	90.42	95.21	92.96	91.53	90.19
OA' (%)	−0.62	−1.37	−1.66	−2.01	−1.17	−2.42	−3.63	−5.37	−1.03	−2.9	−4.47	−5.53
AA' (%)	−0.67	−1.55	−1.68	−1.79	−1.36	−2.37	−3.48	−5.41	−1.36	−3.07	−4.56	−6.92
Kappa'	−0.92	−1.95	−2.21	−2.22	−1.31	−2.58	−3.76	−5.82	−1.03	−3.28	−4.71	−6.05

MDGCN

η_p	Gaussian				Impulsive				Poisson			
	0.1	0.2	0.3	0.4	0.1	0.2	0.3	0.4	0.1	0.2	0.3	0.4
OA (%)	94.82	94.02	93.67	92.81	94.36	92.70	90.62	89.32	94.56	92.73	90.52	89.23
AA (%)	92.61	92.79	92.54	91.79	92.22	90.68	88.57	87.31	91.37	90.67	88.17	86.89
Kappa	93.37	92.50	92.39	91.72	92.30	91.44	89.49	88.16	93.21	91.39	90.24	87.17
OA' (%)	−0.86	−1.66	−2.01	−2.87	−1.32	−2.98	−5.06	−6.36	−1.12	−2.95	−5.16	−6.45
AA' (%)	−0.54	−0.36	−0.61	−1.36	−0.93	−2.47	−4.58	−5.84	−1.78	−2.48	−4.98	−6.26
Kappa'	−0.88	−1.75	−1.86	−2.53	−1.95	−2.81	−4.76	−6.09	−1.04	−2.86	−4.01	−7.08

MARP

η_p	Gaussian				Impulsive				Poisson			
	0.1	0.2	0.3	0.4	0.1	0.2	0.3	0.4	0.1	0.2	0.3	0.4
OA (%)	96.42	95.83	95.31	94.62	95.31	93.87	92.68	90.75	96.21	94.76	93.26	91.73

(continued)

Table 4.9 (continued)

MARP												
η_p	Gaussian				Impulsive				Poisson			
	0.1	0.2	0.3	0.4	0.1	0.2	0.3	0.4	0.1	0.2	0.3	0.4
AA (%)	96.75	96.02	95.78	95.22	95.62	94.12	92.82	91.62	96.37	94.69	93.59	91.42
Kappa	97.17	96.61	95.83	95.69	95.79	94.30	93.11	91.78	96.44	94.83	93.62	91.82
OA' (%)	−0.76	−1.35	−1.87	−2.56	−1.87	−3.31	−4.50	−6.43	−0.97	−2.42	−3.92	−5.45
AA' (%)	−0.61	−1.34	−1.58	−2.14	−1.74	−3.24	−4.54	−5.74	−0.99	−2.67	−3.77	−5.94
Kappa'	−0.52	−1.08	−1.86	−2.00	−1.90	−3.39	−4.58	−5.91	−1.25	−2.86	−4.07	−5.87

SAGE-A achieves good classification results: the OA, AA, and Kappa coefficient of SAGE-A are 96.19, 96.18, and 96.24 respectively, which are better than those of MDGCN and second only to those of MARP. SAGE-A and MARP propose new aggregation methods for the GNN, which overcome the inherent shortcomings of the GCN method to a certain extent, and the HSI classification accuracy is improved. It is also noted that deep CNN-based methods (DR-CNN and CRNN) have not gained absolute advantage compared with traditional machine learning methods (e.g. JSDF and MBCUT), which is different from our inherent impression.

The classification results on Salinas are shown in Table 4.7. MARP still achieves a good classification effect since the adaptive receptive path aggregation can extract useful features more effectively. The OA, AA, and Kappa coefficient of MARP are 97.81, 98.04, and 97.12 respectively. In addition, JSDF achieved better classification results in most classes. The AA of MDGCN is 93.15, which is better than MARP. However, the performances on OA and Kappa of the two methods are lower than MARP since there is an imbalance among different class classifications. It should be emphasized that MDGCN, MBCUT, and MARP demonstrate good classification results since the three methods employ multiscale technology. The results show that multiscale technology can improve the accuracy of hyperspectral classification.

The classification of large HSIs, e.g. Houston 2013, is a challenge to classification algorithms. The results of all investigated methods, shown in Table 4.8, measured by four indicators (i.e. PA, OA, AA, and Kappa), are decline. This is because large images need to identify more details and requre more computational power. In addition, due to the design of the adaptive receptive field, MARP has better adaptability to different classification targets.

(2) *Visual Results*

To intuitively observe the classification results of each method, the classification results on the aforementioned HSI datasets of investigated methods are shown in Figs. 4.8, 4.9, and 4.10. Note that different colours in the same image represent different land scops. As shown in Fig. 4.8, the visualization result of our MARP method on the Pavia University dataset is the most similar to the ground truth. In terms of Salinas and Houston 2013 (i.e., Figs. 4.9 and 4.10) MARP demonstrates the best visualization which is in complete agreement with the ground truths. However, there are many classification mistakes and noise in the classification maps of the comparison methods. Generally, the results show the effectiveness and superiority of our results.

4.4.4 Parament Analysis

The segmentation scale N, number of MARP layers L, learning rate lr and number of iterations T have a great influence on hyperspectral classification performance. In the experiment, a grid search strategy is adopted to find the optimal setting for the

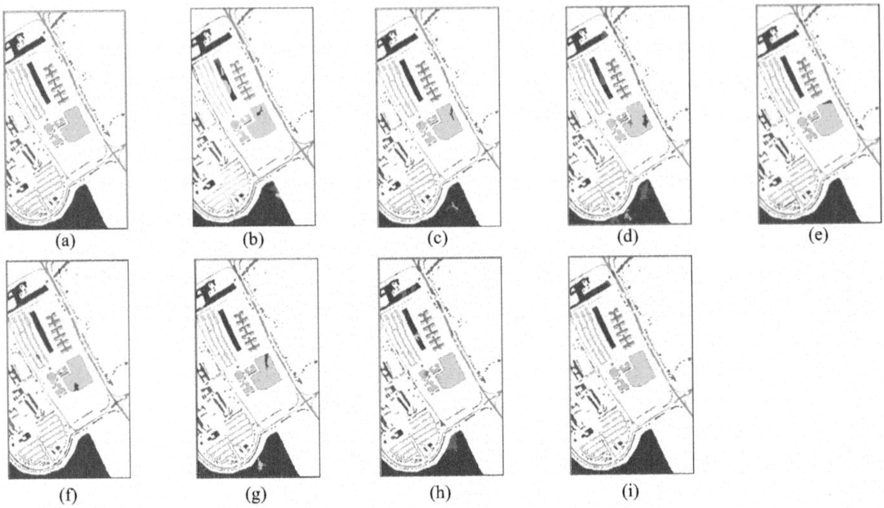

Fig. 4.8 Classification visualization with thirty labelled training pixels per class on Pavia University dataset. **a** Ground truth map; **b** CNN-PPF; **c** DR-CNN; **d** CRNN; **e** SAGE-A; **f** MDGCN; **g** JSDF; **h** MBCUT; **i** MARP

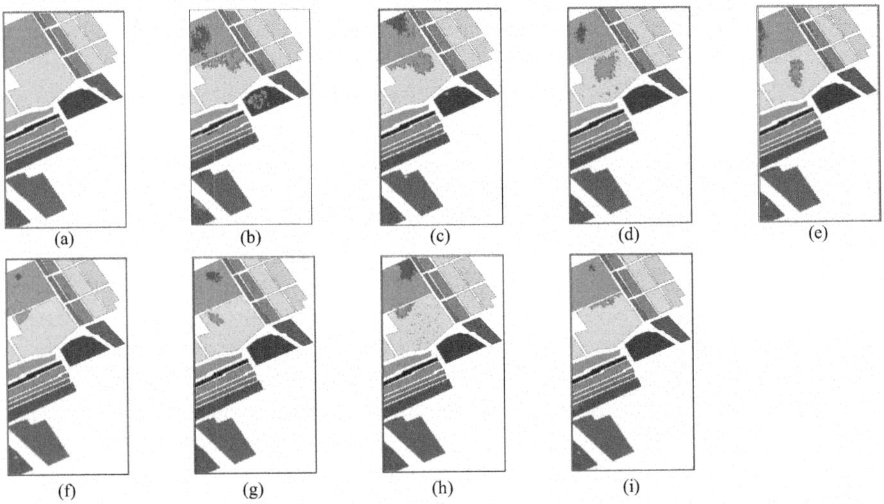

Fig. 4.9 Classification visualization with thirty labelled training pixels per class on Salinas dataset. **a** Ground truth map; **b** CNN-PPF; **c** DR-CNN; **d** CRNN; **e** SAGE-A; **f** MDGCN; **g** JSDF; **h** MBCUT; **i** MARP

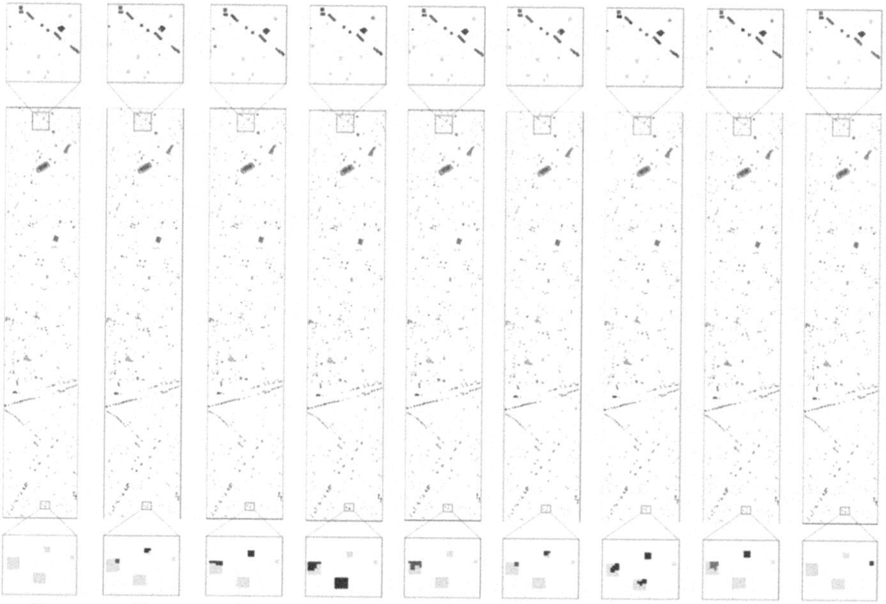

Fig. 4.10 Classification visualization with thirty labelled training pixels per class on Houston 2013 dataset. **a** Ground truth map; **b** CNN-PPF; **c** DR-CNN; **d** CRNN; **e** SAGE-A; **f** MDGCN; **g** JSDF; **h** MBCUT; **i** MARP

parameters. We divide our parameters into two groups and their classification effects are measured by OA.

(a) *Impact of L and N*

In the experiment, the impacts of L and N are investigated in detail. L is set in the range of [1, 6] for the three datasets. In addition, N is varied from 4000 to 16,000 at an interval of 2000 for the Pavia University dataset, 400 to 1600 at an interval of 200 for the Salinas dataset, and 5000 to 35,000 for the Houston dataset. The results are demonstrated in Fig. 4.11. Both parameters have a great influence on classification accuracy. A further tendency can be observed: with an increase in segment scales, the classification accuracy also improves because the greater the number of segmentations, the more precise the hyperspectral image segmentation. However, the larger N is, the more nodes are processed in the subsequent network, and the computational complexity increases exponentially. In terms of L, we conclude that deep neural networks can extract deep features and improve classification accuracy. More importantly, the classification accuracy does not decrease increasing of L, which shows that the algorithm can reduce the oversmoothing problem. However, when L is too large, the method will need more calculation power. To consider the classification accuracy and efficiency, L and N are set shown in Table 4.4.

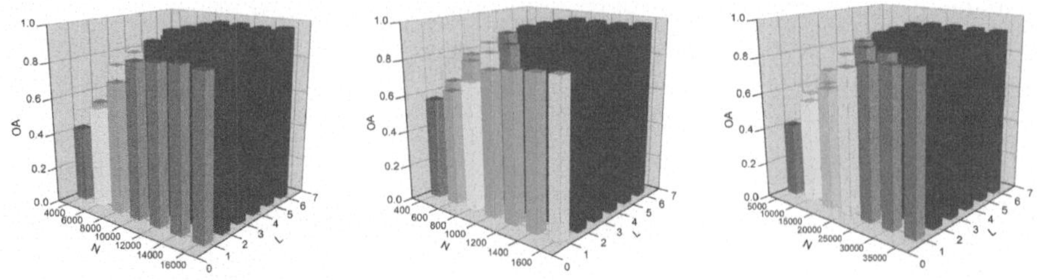

(a) (b) (c)

Fig. 4.11 Parametric sensitivity of L and N. **a** University of Pavia dataset. **b** Salinas dataset. **c** Houston 2013 dataset

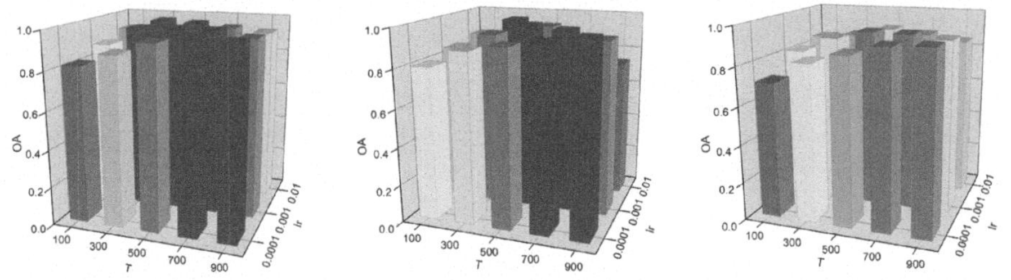

(a) (b) (c)

Fig. 4.12 Parametric sensitivity of *lr* and *T*. **a** University of Pavia dataset. **b** Salinas dataset. **c** Houston 2013 dataset

(b) *Impact of lr and T*

In this experiment, we evaluate the sensitivity of the classification performances to *lr* and *T* in detail. we let *T* vary from 100 to 900 with an interval of 100, and set *lr* as 0.01, 0.001, and 0.001. The results are demonstrated in Fig. 4.12. From the result, we can observe that a large *lr* can train the network quickly. However, the algorithm cannot easily find the most effective result, and the result is unstable. Although a small *lr* can effectively find the optimal solution, the network learning speed is slow. We set $lr = 0.001$, which considers both learning efficiency and classification accuracy. In addition, the appropriate size of *T* is very important to the classification results. Because *T* is too small, the network training is insufficient, and when *T* is too large, it is easy to overfit. We set $T = 500$, when the best result is reached.

4.4.5 Ablation Study

To shed light on the contributions of multiscale and ARP manipulation, in this experiment, the training pixel number is varied from 5 to 30 with an interval of 5 for the three HSI datasets. *S* is set as 1, 2, and 3, and the corresponding algorithms are

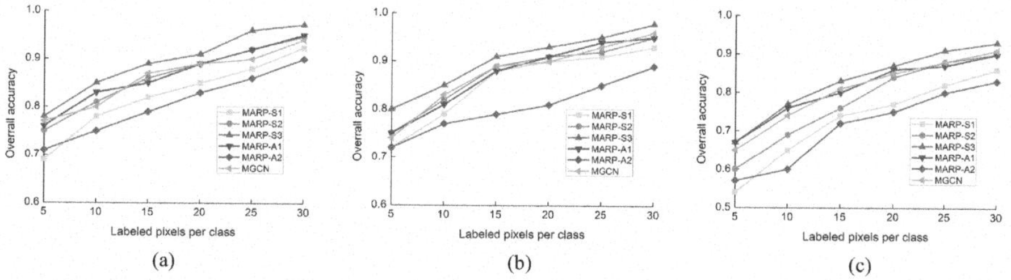

Fig. 4.13 The ablation performances of MARP. **a** Pavia University dataset. **b** Salinas dataset. **c** Houston 2013 dataset

called "MARP-S1," "MARP-S2," and "MARP-S3". "MARP-A1" and "MARP-A2" represent the reduced model by removing the GAT and LSTM, respectively. We adopt "MGCN" to represent the reduced model by removing the ARP manipulation structure and adopting GCN instead. The results are shown in Fig. 4.13. As demonstrated in the results, the classification performances of the four algorithms improve significantly as the labelled pixels increase because more labelled training dates are attributed to a higher classification accuracy rate. Meanwhile, the larger scale model can achieve better classification results, since multiscale can extract more multiscale features from the graph. The experiment indicates that multiscale can effectively improve the accuracy of hyperspectral classification. MARP-S3 achieves a good performance compared with MGCN. It is also important to note that ARP plays a significant role in HSI classification.

4.4.6 Impact of Number of Labelled Examples

In this experiment, we investigate the classification performances of the seven algorithms with different numbers of labelled examples (i.e. pixels) for training. The number of labelled examples per class are varied from 5 to 30 with an interval of 5, and the OA performance acquired by seven algorithms on University of Pavia, Salina, and Houston 2013 datasets are demonstrated in Fig. 4.14. From the results, we find that the proposed method does not obtain good results under 5 and 10 labelled pixels, mainly because ARP does not have enough information to extract a reasonable aggregation path. However, the classification accuracy of our proposed model increases rapidly with an increase in the number of training samples. As the labelled sample number is 30, the proposed MARRP is better than the comparison methods, which proves the superiority of our method.

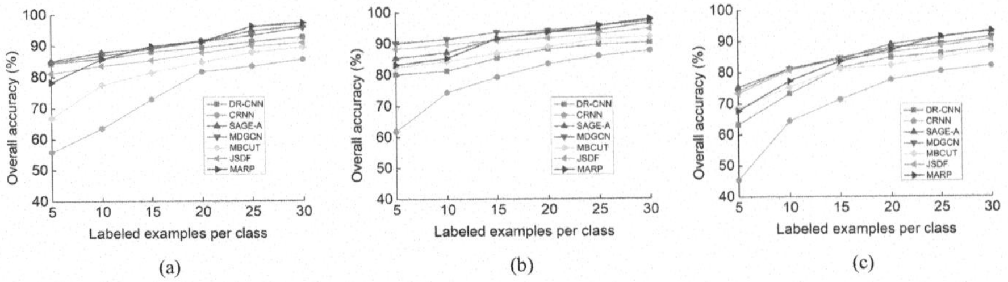

Fig. 4.14 The performances of investigated methods with different numbers of labelled examples per-class. **a** University of Pavia dataset. **b** Salinas dataset. **c** Houston 2013 dataset

4.4.7 The Robustness of the Method Against Noise

To analyse the robustness of GNN-based models (SAGE-A, the MDGCN, and MARP) against different noise distributions, we contaminate the Pavia University test dataset using Gaussian, impulsive, and Poisson noise. The models are trained on noncontaminated samples. We randomly pick $(\eta_p|\varphi|)$ test pixels for contamination, where $|\varphi|$ is the value of noise, and $\eta_p \in \{0.1,0.2,0.3,0.4\}$. In Table 4.9, we gather the results (OA, AA, and Kappa values) obtained in the test sets contaminated with Gaussian, impulsive, and Poisson noise with zero means and various standard deviations. Furthermore, the differences between the classification results (OA', AA', and Kappa') obtained on the original (uncontaminated) test set and the noisy test set are presented. Therefore, we verify the impact of noise with varying intensities on the generalization abilities of all investigated GNNs. The results show that the injection of stronger noise has a negative impact on the performance of all models, and we can observe that their classification performances are significantly reduced. In addition, different models are robust against different types of noise. In Table 4.9, we can see that SAGE-A manifests the best robustness against impulsive noise, whereas it is outperformed by MARP once Passion noise is injected into the test data. Finally, the results show that the proposed model achieves the best classification results no matter what noise is injected. This result indicates that the proposed models have good performance in the face of different types of noise and verifies the superiority of the MARP algorithm design for HSI classification.

4.4.8 Computational Complexity Analysis

To further evaluate the superiority of our MARP in terms of efficiency, the computational complexity is analysed in this part. The computational complexity of our MARP method mainly consists of nodes and edges. Denote by D the spectral dimension of node, α the number of nonzero entries of the adjacency matrix A, and H_1, H_2, H_3, and H_4 the hidden dimension of four convolution layers. The computational complexity of graph convolution is $\mathcal{O}[\alpha(DH_1 + H_1H_2 + H_2H_3 + H_3H_4)]$. If

Table 4.10 Training time (in seconds) of different method

Methods	CNN-PPF	DR-CNN	CRNN	SAGE-A	MDGCN	MARP
Indian Pines	3162.65	405.56	4017.45	2793.37	2976.42	2367.92
University of Pavia	527.16	766.8	2967.32	2123.18	2203.36	1879.23
Salinas	5691.87	232.2	3026.19	1058.91	1227.41	1025.79
Houston 2013	11,952.36	642.60	3534.27	2907.51	3029.29	2679.82

our model iterates T times, then, we express the overall computational complexity of MARP as $\mathcal{O}[\alpha(DH_1 + H_1H_2 + H_2H_3 + H_3H_4)T]$. From the expression formula of computational complexity, we conclude that the computational complexity of MARP is mainly determined by the number of nonzero entries of the adjacency matrix A and the hidden dimension of four convolution layers.

Table 4.10 summarizes the computational complexity of training procedures using the investigated deep-learning methods and the proposed MARP. All experiments were carried out using a PC equipped with a single 3.70G Intel i9-10900K CPU. Note that the execution time of MARP is less than that of comparative methods for the training procedures. The reason is that the superpixel method can greatly reduce the number of nodes in the classification map. Therefore, the size of the input data of the former is smaller than that of the latter.

4.5 Conclusion

In this paper, a novel multiadaptive receptive field graph neural framework was proposed for HSI classification. We introduced an adaptive receptive path aggregation mechanism and regraded the convolution feature of each layer as a timing signal and processed it by LSTM. Therefore, local features of each convolution layer can be preserved in LSTM while the more comprehensive feature representation is available from a larger receptive field. More significant, to alleviate the local optimum problem due to adaptive path selection, a multireceptive field framework was designed to explore the multiscale receptive field of the graph. Extensive experiments on three public HSI datasets verified the effectiveness and advantages of the proposed MARP when compared with related state-of-the-art HSI classification methods.

References

1. B. Rasti et al., Feature extraction for hyperspectral imagery: the evolution from shallow to deep: overview and toolbox. IEEE Geosci. Remote Sens. Mag. **8**(4), 60–88 (2020). https://doi.org/10.1109/-MGRS.2020.2979764

2. P. Ghamisi, N. Yokoya, J. Li, W. Liao, S. Liu, J. Plaza, B. Rasti, A. Plaza, Advances in hyperspectral image and signal processing: a comprehensive overview of the state of the art. IEEE Geosci. Remote Sens. Mag. **5**(4), 37–78 (2017)
3. J. Peng, W. Sun, Q. Du, Self-paced joint sparse representation for the classification of hyperspectral images. IEEE Trans. Geosci. Remote Sens. **57**(2), 1183–1194 (2018)
4. X. Lu, B. Wang, X. Zheng, X. Li, Exploring models and data for remote sensing image caption generation. IEEE Trans. Geosci. Remote Sens. **56**(4), 2183–2195 (2018)
5. Z. Zhao, P. Zheng, S. Xu, X. Wu, Object detection with deep learning: a review. IEEE Trans. Neural Netw. Learn. Syst. **30**(11), 3212–3232 (2019)
6. D. Hong, W. He, N. Yokoya, J. Yao, L. Gao, L. Zhang, J. Chanussot, X. Zhu, Interpretable hyperspectral AI: when non-convex modeling meets hyperspectral remote sensing. IEEE Geosci. Remote Sens. Mag. **11**, 4051 (2021). https://doi.org/10.1109/MGRS.2021.3064051
7. D. Hong, L. Gao, N. Yokoya, J. Yao, J. Chanussot, Q. Du, B. Zhang, More diverse means better: multimodal deep learning meets remote-sensing imagery classification. IEEE Geosci. Remote Sens. **9**, 16820 (2020). https://doi.org/10.1109/TGRS.2020.3016820
8. Y. Ding, X. Zhao, Z. Zhang, W. Cai, N. Yang, Graph sample and aggregate-attention network for hyperspectral image classification. IEEE Geosci. Remote Sens. Lett. **19**, 1–5 (2021). https://doi.org/10.1109/LGRS.2021.306-2944
9. D. Hong, N. Yokoya, J. Chanussot, X. Zhu, An augmented linear mixing model to address spectral variability for hyperspectral unmixing. IEEE Trans. Image Process. **28**(4), 1923–1938 (2019)
10. S. Wan, C. Gong, P. Zhong, S. Pan, G. Li, J. Yang, Hyperspectral image classification with context-aware dynamic graph convolutional network. IEEE Trans. Geosci. Remote Sens. **59**(1), 597–612 (2021)
11. P. Zhong, Z. Gong, J. Shan, Multiple instance learning for multiple diverse hyperspectral target characterizations. IEEE Trans. Neural Netw. Learn. Syst. **14**, 1–13 (2019)
12. J. Feng et al., Generative adversarial networks based on collaborative learning and attention mechanism for hyperspectral image classification. Remote Sens. **12**(7), 1149 (2020)
13. M. Pesaresi, A. Gerhardinger, F. Kayitakire, A robust built-up area presence index by anisotropic rotation-invariant textural measure. IEEE J. Select. Top. Appl. Earth Observ. Remote Sens. **1**(3), 180–192 (2008)
14. K. Djerriri, A. Safia, R. Adjoudj, M.S. Karoui, Improving hyperspec-tral image classification by combining spectral and multiband compact texture features, in *Proceedings of the IEEE International Geosciences Remote Sensing Symposium (IGARSS)* (2019), pp. 465–468
15. W. Li, G. Wu, F. Zhang, Q. Du, Hyperspectral image classification using deep pixel-pair features. IEEE Trans. Geosci. Remote Sens. **55**(2), 844–853 (2017)
16. C. Liu, J. Li, L. He, A. Plaza, S. Li, B. Li, Naive gabor networks for hyperspectral image classification. IEEE Trans. Neural Netw. Learn. Syst. **32**(1), 376–390 (2020)
17. Y. Cai, X. Liu, Z. Cai, BS-Nets: an end-to-end framework for band selection of hyperspectral image. IEEE Geosci. Remote Sens. **58**, 1969–1984 (2020)
18. Y. Chen, Z. Lin, X. Zhao, G. Wang, Y. Gu, Deep learning-based classification of hyperspectral data. IEEE J. .Select Top. Appl. Earth Observ. Remote Sens. **7**(6), 2094–2107 (2014)
19. Y. Chen, X. Zhao, X. Jia, Spectral–spatial classification of hyperspectral data based on deep belief network. IEEE J. Select. Top. Appl. Earth Observ. Remote Sens. **8**(6), 2381–2392 (2015)
20. L. Mou, P. Ghamisi, X.X. Zhu, Deep recurrent neural networks for hyperspectral image classification. IEEE Trans. Geosci. Remote Sens. **55**(7), 3639–3655 (2017)
21. M. Zhang, W. Li, Q. Du, Diverse region-based CNN for hyperspectral image classification. IEEE Trans. Image Process. **27**(6), 2623–2634 (2018)
22. T. Li, J. Zhang, Y. Zhang, Classification of hyperspectral image based on deep belief networks, in *Proceedings of the IEEE ICIP* (2014), pp. 5132–5136
23. R. Hang, Q. Liu, D. Hong, P. Ghamisi, Cascaded recurrent neural networks for hyperspectral image classification. IEEE Geosci. Remote Sens. **57**(8), 5384–5394 (2019). https://doi.org/10.1109/TGRS.2019.2899129

24. W. Hu, Y. Huang, L. Wei, F. Zhang, H. Li, Deep convolutional neural networks for hyperspectral image classification. J. Sens. **2015**, 1–12 (2015)
25. J. Yang, Y. Zhao, J.C. Chan, C. Yi, Hyperspectral image classification using two-channel deep convolutional neural network, in *Proceedings of the IEEE IGARSS* (2016), pp. 5079–5082
26. Y. Li, H. Zhang, Q. Shen, Spectral-spatial classification of hyperspectral imagery with 3D convolutional neural network. Remote Sens. **9**(1), 67 (2017)
27. D. Hong, N. Yokoya, G. Xia, J. Chanussot, X. Zhu, X-ModalNet: a semi-supervised deep cross-modal network for classification of remote sensing data. ISPRS J. Photogram. Remote Sens. **167**, 12–23 (2020)
28. C. Wang, S. Pan, R. Hu, G. Long, J. Jiang, C. Zhang, Attributed graph clustering: a deep attentional embedding approach. in *Proceedings of the International Joint Conference on Artificial Intelligent (IJCAI)* (2019), pp. 3670–3676
29. D. Hong, N. Yokoya, J. Chanussot, J. Xu, X. Zhu, Learning to propagate labels on graphs: an iterative multitask regression framework for semi-supervised hyperspectral dimensionality reduction. ISPRS J. Photogram. Remote Sens **158**, 35–49 (2019)
30. Z. Zhang, P. Cui, W. Zhu, Deep learning on graphs: a survey. IEEE Trans. Knowl. Data Eng. **32**, 1–20 (2020)
31. T.N. Kipf, M. Welling, Semi-supervised classification with graph convolutional networks, in *Proceedings of the ICLR* (2017)
32. Z. Zhang, P. Cui, W. Zhu, Deep learning on graphs: a survey. IEEE Trans. Knowl. Data Eng. **45**, 1–20 (2020)
33. A. Qin, Z. Shang, J. Tian, Y. Wang, T. Zhang, Y.Y. Tang, Spectral–spatial graph convolutional networks for semisupervised hyperspectral image classification. IEEE Geosci. Remote Sens. Lett. **16**(2), 241–245 (2019). https://doi.org/10.1109/LGRS.2018.2869563
34. A. Sha, B. Wang, X. Wu, L. Zhang, Semisupervised classification for hyperspectral images using graph attention networks. IEEE Geosci. Remote Sens. Lett. **8**, 23 (2020)
35. S. Wan, C. Gong, P. Zhong, B. Du, L. Zhang, J. Yang, Multiscale dynamic graph convolutional network for hyperspectral image classification. IEEE Trans. Geosci. Remote Sens. **58**(5), 3162–3177 (2020)
36. Y. Ding, X. Zhao, Z. Zhang, W. Cai, N. Yang, Multi-scale graph sample and aggregate network with context-aware learning for hyperspectral image classification. IEEE J. Select. Top. Appl. Earth Observ. Remote Sens. **14**, 4561–4572 (2021). https://doi.org/10.1109/JSTARS.2021.3074469
37. D. Hong, L. Gao, J. Yao, B. Zhang, A. Plaza, J. Chanussot, Graph convolutional networks for hyperspectral image classification. IEEE Trans. Geosci. Remote Sens. **59**, 5966–5978 (2020). https://doi.org/10.1109/TGRS.2020.3015157
38. T. Kipf, M. Welling, Semi-supervised classification with graph convolutional networks, in *International Conference on Learning Representations* (2017)
39. J. Bruna, W. Zaremba, A. Szlam, Y. LeCun, Spectral networks and locally connected networks on graphs, in *Proceedings of the International Conference on Learning Representive (CLR)* (2014), pp. 1–14
40. A. Sandryhaila, J.M.F. Moura, Discrete signal processing on graphs. IEEE Trans. Signal Process. **61**(7), 1644–1656 (2013)
41. D.I. Shuman, S.K. Narang, P. Frossard, A. Ortega, P. Vandergheynst, The emerging field of signal processing on graphs: Extending high-dimensional data analysis to networks and other irregular domains. IEEE Sig. Process. Mag. **30**(3), 83–98 (2013)
42. P.A. Gagniuc, *Markov Chains: From Theory to Implementation and Experimentation* (Wiley, 2017)
43. S. Bernhard, A. Smola, and K-R. Müller. Nonlinear component analysis as a kernel eigenvalue problem. Neural comput **10**(5):1299–1319 (1998)
44. R. Achanta, A. Shaji, K. Smith, A. Lucchi, P. Fua, S. Strunk, Slic superpixels compared to state-of-the-art superpixel methods. IEEE Trans. Pattern Anal. Mach. Intell. **34**(11), 2274–2282 (2012)

45. Y. Cai, Z. Zhang, Z. Cai, X. Liu, X. Jiang, Q. Yan, Graph convolutional subspace clustering: a robust subspace clustering framework for hyperspectral image. IEEE Trans. Geosci. Remote Sens. **15**, 1–12 (2020)
46. N. Srivastava, G. Hinton, A. Krizhevsky, I. Sutskever, R. Salakhutdinov, Dropout: a simple way to prevent neural networks from overfitting. J. Mach. Learn. Res. **15**(1), 1929–1958 (2014)
47. S. Zhang, S. Li, Spectral-spatial classification of hyperspectral images via multiscale superpixels based sparse representation, in *Proceedings of the IEEE IGARSS* (2016), pp. 2423–2426

Chapter 5
Graph Neural Network via Edge Convolution for Hyperspectral Image Classification

5.1 Introduction

Hyperspectral images have been widely used in many remote sensing applications due to their rich spectral and spatial information, and up to now, a wide range of applications have been benefited from the development of this direction, including urban development, land monitoring, scene interpretation, and resource exploration 0. Among these applications, HSI classification is a common technique that facilitates the study of the chemical properties of scene materials remotely.

Over the past few decades, diverse kinds of approaches have been proposed for HSI classification. The early-staged methods are mainly based on conventional pattern recognition methods, such as K-nearest neighbor classifier [1], support vector machines (SVM) [2, 3], and sparse representation [4] However, the performance of these HSI classification methods is heavily dependent on the handcrafted features, which heavily rely on professional expertise and are pretty empirical [5].

Deep learning provides an ideal solution to obtaining more accurate classification by automatically extract discriminative features [6–8]. Many CNN-based variants have been designed for HSI classification from the 1D-CNN to the 3D-CNN [9], from the single CNN to the hybrid CNN [10–12]. Although the existing CNN-based methods have achieved good performance to some extent, they need many training parameters and tend to be overfitting due to the scarcity of the training samples. Meanwhile, CNN tends to blur the classification boundary using a fixed shape kernel around the central pixel to extract features 0. Therefore, the accurate classification of HSI is still challenging.

In recent years, GNN-based methods have been proposed for HSI classification. For example, Qin et al. [13] first apply graph convolutional network (GCN) to HSI classification by making full use of spectral-spatial neighbor information. Mou et al. [14] uses a learnable graph convolutional network (GCN) to obtain a nonlocal, data-driven graph representation. Hong et al. [15] systematically analyze CNN and GCN with a focus on HSI classification and fuse the features extracted by CNN and GCN.

Y. Ding et al., *Graph Neural Network for Feature Extraction and Classification of Hyperspectral Remote Sensing Images*, Intelligent Perception and Information Processing, https://doi.org/10.1007/978-981-97-8009-9_5

Wan et al. [16] utilize multi-scale graph convolution to exploit the spatial information extensively. After that, inductive learning and attention mechanism are applied in GCN for HSI classification [5, 17–19].

GCN generates point features directly from neighbors' embeddings, while it inevitably aggregates the information of neighbors that actually belong to another category, leading to the loss of some discriminative information. Comparatively, EdgeConv [20] can relieve the problem of GCN to some extent since it aggregates information according to the relationships between a point and its neighbors. Inspired by the wide application of EdgeConv in the task of point cloud segmentation, we try to explore its potential for HSI procession. The main contributions of this article are as follows:

- We introduce a GNN structure where the message passing is followed by the EdgeConv layer. Within the proposed graph structure, our model can adaptively capture the inter-relationship of the representative descriptors and fully exploit the discriminative features on graph.
- We propose a novel end-to-end network architecture that combines GNN and CNN for superpixel and pixel-level feature fusion without loss of resolution, which is achieved by incorporating graph encode and decode operations.
- Experiments on three benchmark datasets show that our method outperforms other compared HSI classification algorithms.

The rest of the paper is organized as follows. Section 5.2 briefly reviews the methodology of GNN and EdgeConv. Then, the proposed method is introduced in Sect. 5.3. In Sect. 5.4, experiments on two real HSI datasets are conducted to evaluate the proposed method. Conclusion is followed in Sect. 5.5.

5.2 Related Work

This section reviews the general framework of GNN and its example (EdgeConv) because they are the fundamental work of our proposed method. The differences between EdgeConv and GCN are also illustrated in this section.

5.2.1 GNN

Let $\mathcal{G} = (\mathcal{V}, \mathcal{E})$ represents the given undirected graph, where $\mathcal{V}$ is the set of vertices and $\mathcal{E}$ is the set of edges representing the connectivity between vertices $v \in \mathcal{V}$. Graph neural network propagates information according to message passing mechanism, which commonly consists of message aggregation phase and update phase. During the message passing phase [21], hidden states h_v^l at each node in the graph are updated based on message m_v^{l+1} according to

$$m_v^{l+1} = \sum_{w \in N(v)} M\left(h_v^l, h_w^l, e_{vw}\right), \tag{5.1}$$

$$Ah_v^{l+1} = U\left(h_v^l, m_v^{l+1}\right), \tag{5.2}$$

where M is the vertex feature aggregation and U is the vertex messageupdate function, $N(v)$ denotes the neighbors of v in the graph $\mathcal{G}$, h_v^l and h_v^{l+1} are the vertex features at the l-th layer and $(l+1)$-th layer, respectively.

5.2.2 *EdgeConv*

The recent EdgeConv method demonstrates the potential of GCN for point cloud-related applications and beats the state-of-the-art in the task of point cloud segmentation. Instead of aggregating neighborhood features directly, EdgeConv proposes to first get local neighborhood information for each neighbor by subtracting the feature of the central vertex from its own feature, which has the properties lying between translation-invariance and non-locality. The message passing of EdgeConv [22] can be represented as

$$Ah_v^{l+1} = max\{MLP\left(concat\left(h_v^l, h_w^l - h_v^l\right)\right)|w \in N(v)\}, \tag{5.3}$$

With both $h_w^t - h_v^t$ and h_v^t as input, the model considers the local topological structure while keeping global spectral information. Figure 5.1 illustrates the corresponding message passing process of GCN and EdgeConv, respectively.

5.3 Proposed Method

Let $\mathbf{X} \in \mathbb{R}^{H \times W \times B}$ denote the hyperspectral dataset, where H, W and B represent the height, width, and the number of bands, respectively. The task for hyperspectral classification aims to predict the labels of all the pixels in the HSI by a classification

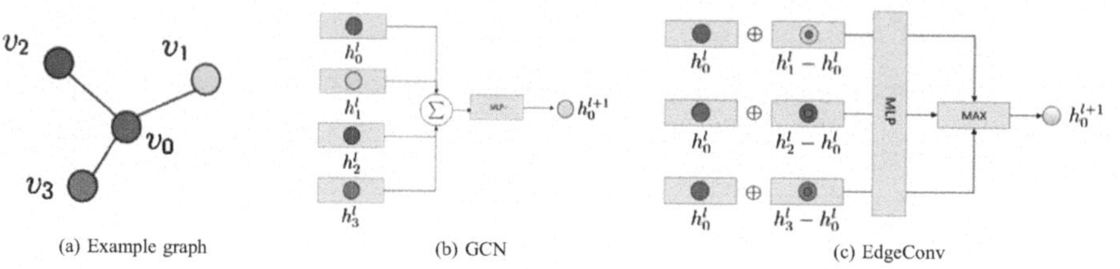

Fig. 5.1 **a** Gives an example graph with v_0 as the central node. **b** and **c** Describe the corresponding message passing process of GCN and EdgeConv respectively

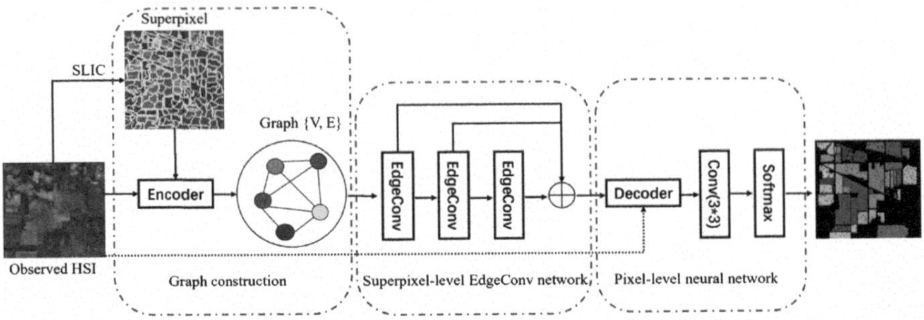

Fig. 5.2 Schematic depiction of the proposed method. Firstly, a superpixel map is generated based on the input image and used to produce the association matrix $\mathbf{Q}$. After encoder, the HSI is converted to an undirected graph $\mathcal{G} = (\mathcal{V}, \mathcal{E})$ that models the relationship between different regions in HSI by treating each superpixel as a graph node. In EdgeConv network, the graph is fed into a three-layer to obtain the high-level feature map. And then, the corresponding coarse segmentation result $\mathbf{X}^*$ is obtained after decoder. Finally, the refined feature map is achieved through a pixel-level neural network for the final classification

model given the supervised data set. We propose an EdgeConv-based graph neural network for HSI classification model dubbed EGNN, as shown in Fig. 5.2, which consists of three main blocks: graph construction block, an EdgeConv block, and a CNN prediction block. These three blocks are connected by encode and decode operations.

5.3.1 Graph Construction

To reduce the computational complexity and meanwhile preserve the local structure of HSI, the GNN-based models usually work on superpixel-based nodes but not the pixel-based nodes. Similarly, we first apply the simple linear iterative clustering (SLIC) method to partition the entire image into many spatially connected superpixels. By treating each superpixel as a graph node and establishing the adjacency relations between superpixels, the HSI is converted to an undirected graph $\mathcal{G} = (\mathcal{V}, \mathcal{E})$. The set of vertices $\mathcal{V}$ here denotes the group of obtained superpixels, which can be represented by a feature matrix $\mathbf{V} \in \mathbb{R}^{N \times B}$ which is defined by the average spectral signatures of each superpixel, where N is the number of superpixels. $\mathcal{E}$ denotes the edges between the superpixels, which can be represented by a spatial adjacency matrix. Specifically, the adjacency matrix $\mathbf{A} \in \mathbb{R}^{N \times N}$ of graph $\mathcal{G}$ is defined as

$$\mathbf{A}_{i,j} = \begin{cases} 1, \text{ if } \mathcal{S}_i \text{ and } \mathcal{S}_j \text{ are adjacency}, \\ 0, \text{ otherwise}, \end{cases} \tag{5.4}$$

where $\mathcal{S}_i$ is the i-th superpixel that consists of several homogeneous pixels $\{\mathbf{x}_1^i, \mathbf{x}_2^i, \cdots, \mathbf{x}_{n_i}^i\}$, n_i is the number of pixels contained in $\mathcal{S}_i$.

We use the graph encoder and decoder to propagate features from image pixels to graph nodes and transform graph features back to the image space, respectively. Let $\mathbf{Q} \in \mathbb{R}^{HW \times N}$ represent the association matrix introduced by SLIC, which is defined as

$$Q_{i,j} = \begin{cases} 1, \text{ if } \mathbf{x}_i \in \mathcal{S}_j, \\ 0, \text{ otherwise.} \end{cases} \tag{5.5}$$

Therefore, we can encode HSI into graph nodes by the form of matrix multiplication as follows

$$\mathbf{V} = \text{Encode}(\mathbf{X}; \mathbf{Q}) = \widehat{\mathbf{Q}}^T \text{Flatten}(\mathbf{X}), \tag{5.6}$$

where $\widehat{\mathbf{Q}}$ is the normalized $\mathbf{Q}$ by column, i.e., $\widehat{\mathbf{Q}}_{i,j} = \mathbf{Q}_{i,j} / \sum_m \mathbf{Q}_{m,j}$. Flatten$(\cdot)$ denotes flattening the HSI data by the spatial dimension. The descriptions about Encoder is illustrated in Sect. 5.3.3.

5.3.2 *Superpixel-Level EdgeConv Network*

With the obtained graph $\mathcal{G}$, inference can then be performed by passing messages between nodes and along the edges that connect them. The edge messages here are defined as $e_{ij} = h_\Theta(\mathbf{x}_i, \mathbf{x}_j)$, where $h_\Theta : \mathbb{R}^F \times \mathbb{R}^F \to \mathbb{R}^F$. The EdgeConv operation takes a tensor $\mathbf{V}$ of shape $N \times F$ as input, computes a tensor $\widetilde{\mathbf{V}}$ of shape $N \times F$ after pooling among neighboring edge features obtained from a multi-layer perceptron. Shortcut connections are used to include all the EdgeConv outputs to aggregate multi-scale features with a tensor of shape $N \times 3F$. Dropout with probability of 0.5 and ReLU are used in each EdgeConv layer.

5.3.3 *Pixel-Level CNN Network*

In order to assign specific label for each pixel, we need to propagate the features to pixels from superpixel-based nodes, The decode operation takes the inputs of transformed vertex features $\widetilde{\mathbf{V}}$ and the assignment matrix $\mathbf{Q}$, and produces the corresponding 3D feature map $\mathbf{X}^*$. which is achieved by the decode operation

$$\mathbf{X}^* = \text{Decode}\left(\widetilde{\mathbf{V}}; \mathbf{Q}\right) = \text{Reshape}\left(\mathbf{Q}\,\widetilde{\mathbf{V}}\right). \tag{5.7}$$

where Reshape$(\cdot)$ denotes restoring the spatial dimension of the flattened data. Since the encode and decode architecture will progressively reduce the spatial resolution, the samples are further fed into the 3×3 convolution layer owing to its ability to extract local spectral-spatial features from HSI. After a Softmax layer, it outputs per-pixel classification scores for C semantic label.

In our network, a cross-entropy error over all labeled pixels is used as the loss function, namely

$$\mathcal{L} = - \sum_{s \in \mathbf{Y}_{labeled}} \sum_{f=1}^{C} \mathbf{Y}_{sf} \ln \mathbf{O}_{sf}, \tag{5.8}$$

where $\mathbf{Y}$ represents the label matrix, $\mathbf{O}_{sf}$ denotes the probability of the s-th pixel belonging to the f-th class, and C is the number of object classes.

5.3.4 *Time Complexity Analysis*

In this subsection, we only conduct the time complexity analysis for the EdgeConv block since the time cost of graph construction block and CNN prediction block are fixed and thoroughly researched in many literatures. In our method, each node is associated with an F-dimensional feature vector. Let $\|\mathbf{A}\|_0$ represent the number of nonzeros in the adjacency matrix. According to the message passing of EdgeConv, computing each embedding requires $O(2kF^2)$ in average due to the multiplication with $W^{(l)}$ in MLP, where k denotes the average degree of each node, so the time complexity for each layer is $O(\|\mathbf{A}\|_0 F^2 + NF^2)$. Thus the overall time complexity per epoch for an L-layer EdgeConv network becomes $O(\|\mathbf{A}\|_0 F^2 L + NF^2 L)$. Considering that N and $\|\mathbf{A}\|_0$ (approximately equal to kN) are much smaller than the size of HSI ($H \times W$), the computational cost can be greatly reduced with the help of superpixel segmentation.

5.4 Experiments

In this section, we conduct experiments to validate the effectiveness of our proposed method EGNN on three real-world benchmark datasets, namely Indian Pines (IP), Pavia University (PU) and Kennedy Space Center (KSC). The Indian Pines dataset is captured over the agricultural test site in the Indian Pines, which contains 200 spectral bands with a size of 145×145 and 16 terrain classes for research and analysis. The Pavia University dataset is the continuous imaging of 115 spectral bands with a size of 610×340. This dataset includes 9 land-cover classes. The KSC data contains 176 bands with a size of 512×614. For classification purpose, 13 classes representing the various land cover types that occur in this environment were defined for the site.

5.4.1 Experimental Settings

To quantitatively evaluate different models for hyperspectral image classification tasks from various aspects, three metrics, including overall accuracy (OA), average accuracy (AA), and kappa coefficient are used to evaluate the performance of all the compared methods. For both datasets, we randomly select 30 labeled pixels in each class for training and choose 15 labeled pixels if the corresponding class has less than 30 samples. Besides, the learning rate and the number of training epochs are set to 0.001 and 2000, respectively. All the reported results are calculated based on the average of ten training sessions to obtain stable results, and the best results are highlighted in bold.

5.4.2 Classification Results

In our experiments, we compare our method with several state-of-the-art HSI classification methods: SVM, Multilayer Perception (MLP), 3DCNN [9], MSDN [23], NLGCN [15] and GSAGE [5] and S^2GCN [13]. The quantitative results on all the three datasets are shown in Tables 5.1, 5.2, and 5.3, respectively. From the results, we can see that GNN-based methods, including GSAGE, S^2GCN and the proposed method, show better performance than traditional machine learning and deep learning models (i.e., SVM, MLP, 3DCNN). Besides, our method significantly outperforms other approaches. Specifically, our method increases accuracies significantly by 7.38% of OA, 4.28% of AA, 8.83% of Kappa compared with the second-best method S^2GCN on IP dataset.

The classification maps of investigated methods on the IP, PU, and KSC datasets are demonstrated in Figs. 5.3, 5.4, and 5.5. From a visual perspective, the classification maps obtained by these comparison methods are consistent with the quantitative results in Tables 4.1, 4.2, and 4.3. From the visible results, we can conclude that classifiers (SVM and MLP), relying only on spectral information, may cause large Salt-and-pepper noise. Although spectral-spatial Classful network (MSDN and 3DCNN) have solved this problem, their classification maps are excessive smoothness. On the contrary, GNN based methods can not only eliminate noisy scattered points in the classification map, but also preserve edge information. The EGNN network proposed in this chapter has the highest classification accuracy and minimizes the noise caused by misclassification in the classification images. From the above results, EGNN is obviously superior to comparative methods (only rely on GNN or CNN) by fusing the features at the super-pixel level and the pixel-level CNN features.

To evaluate the training efficiency of EGNN, we compare its training time with those of other deep learning methods on three datasets, and experimental results are recorded in Table 5.4.

Table 5.1 Classification results (in percent) of different methods on IP real datasets

No	SVM	MLP	3DCNN	MSDN	NLGCN	GSAGE	S^2GCN	EGNN
1	93.75	87.50	100.0	96.77	100.0	100.0	100.0	93.75
2	66.17	58.94	79.54	82.68	75.39	74.11	74.95	82.11
3	64.88	63.88	69.63	59.87	76.88	80.38	84.78	93.00
4	82.61	86.96	96.14	93.79	94.20	98.55	93.07	97.10
5	86.98	84.11	90.29	89.83	91.83	95.58	93.53	97.79
6	88.14	87.14	85.87	94.93	90.57	99.43	96.26	98.71
7	100.0	92.31	100.0	100.0	100.0	100.0	100.0	100.0
8	92.19	92.19	95.98	99.04	100.0	100.0	99.55	100.0
9	100.0	100.0	100.0	100.0	100.0	80.00	100.0	100.0
10	70.49	70.49	82.91	81.47	92.78	83.86	93.81	96.07
11	54.27	57.77	65.20	76.33	73.81	78.80	79.71	91.55
12	56.84	60.57	77.62	86.30	83.84	75.84	93.73	82.42
13	96.57	96.57	99.43	100.0	99.43	100.0	98.82	100.0
14	81.13	83.00	90.61	93.20	91.50	93.60	95.12	99.91
15	63.76	71.07	74.72	81.90	88.48	98.03	87.46	100.0
16	98.41	100.0	98.41	100.0	100.0	100.0	100.0	100.0
OA (%)	69.72	70.07	79.19	83.14	83.83	85.36	87.35	93.12
AA (%)	81.01	80.78	87.88	89.76	91.17	91.13	93.17	95.78
Kappa	66.00	66.31	76.55	80.73	81.70	83.35	85.61	92.14

Table 5.2 Classification results (in percent) of different methods on PU real datasets

No	SVM	MLP	3DCNN	MSDN	NLGCN	GSAGE	S^2GCN	EGNN
1	69.78	73.63	82.25	91.52	86.81	85.91	87.73	96.91
2	73.60	77.98	80.75	85.15	83.22	88.45	88.12	93.62
3	83.91	83.95	94.83	90.96	91.25	91.59	88.08	87.97
4	86.98	91.76	86.85	97.62	96.11	92.02	94.32	94.50
5	99.16	98.63	99.92	100.0	99.62	100.0	99.77	99.92
6	76.66	82.54	77.04	87.41	86.40	89.79	74.83	99.96
7	94.62	91.85	90.23	94.05	96.69	99.23	93.30	99.92
8	82.04	77.17	85.60	95.09	87.46	86.69	96.24	95.37
9	99.67	100.0	98.58	98.91	100.0	99.89	99.89	98.36
OA (%)	77.54	80.59	83.35	89.39	87.11	89.41	89.74	95.31
AA (%)	85.16	86.40	88.45	93.41	91.95	92.62	92.80	96.28
Kappa	71.49	75.25	78.59	86.21	83.35	86.21	87.25	93.84

Table 5.3 Classification results (in percent) of different methods on KSC real datasets

No	SVM	MLP	3DCNN	MSDN	NLGCN	GSAGE	S^2GCN	EGNN
1	88.95	88.29	93.57	96.72	97.13	96.79	97.11	100.0
2	84.37	82.16	74.65	89.20	90.94	91.92	90.38	100.0
3	84.91	84.07	85.40	96.90	94.65	97.92	95.93	97.35
4	54.37	21.22	22.52	54.50	59.46	54.23	66.36	96.85
5	64.96	60.99	84.73	84.74	85.65	84.89	86.51	80.15
6	58.69	46.93	79.90	79.90	75.58	87.04	88.66	98.99
7	91.20	85.60	98.00	94.67	95.20	98.27	100.0	100.0
8	80.32	80.25	69.58	86.78	93.17	97.33	93.43	99.00
9	73.22	90.78	81.63	90.61	96.88	97.14	100.0	100.0
10	85.00	90.99	94.92	99.47	97.99	96.39	98.10	100.0
11	98.53	94.34	99.74	98.46	98.92	100.00	99.48	94.60
12	83.00	83.81	82.47	86.68	92.79	95.03	97.86	96.19
13	98.48	98.83	100.0	99.44	99.79	100.0	100.0	100.0
OA (%)	84.44	84.04	85.87	91.83	93.70	94.70	95.65	98.26
AA (%)	80.46	77.56	82.24	89.08	90.63	92.07	93.37	97.16
Kappa	82.62	82.17	84.22	90.87	92.95	94.08	95.13	98.05

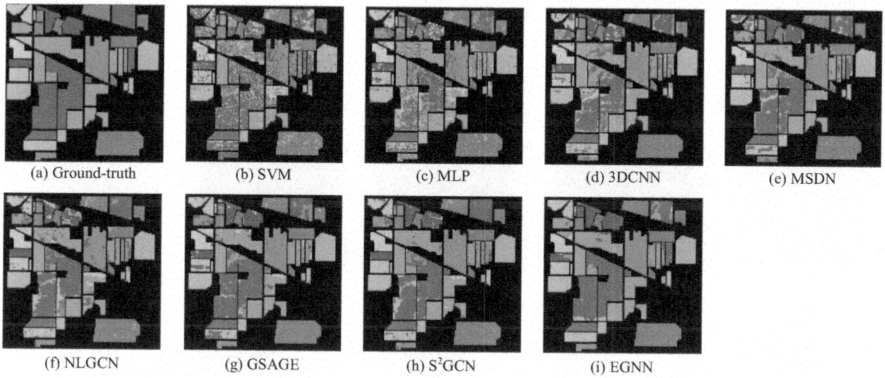

Fig. 5.3 Classification maps of comparison methods on IP dataset

From Table 5.4, it can be seen that CNN based methods require a significant amount of computational time to train the networks due to a large number of parameters. Comparatively, the GNN-based methods significantly reduce the number of parameters, and the running speeds are significantly accelerated by convolving the target node and its neighboring nodes using the constructed graph structure data. Although EGNN reduces the size of graph data by utilizing superpixel segmentation, its runtime is slightly longer compared to NLGCN and GSAGE due to the integration of pixel-level features using CNN. Since only 1D and 2D convolutions are adopted in

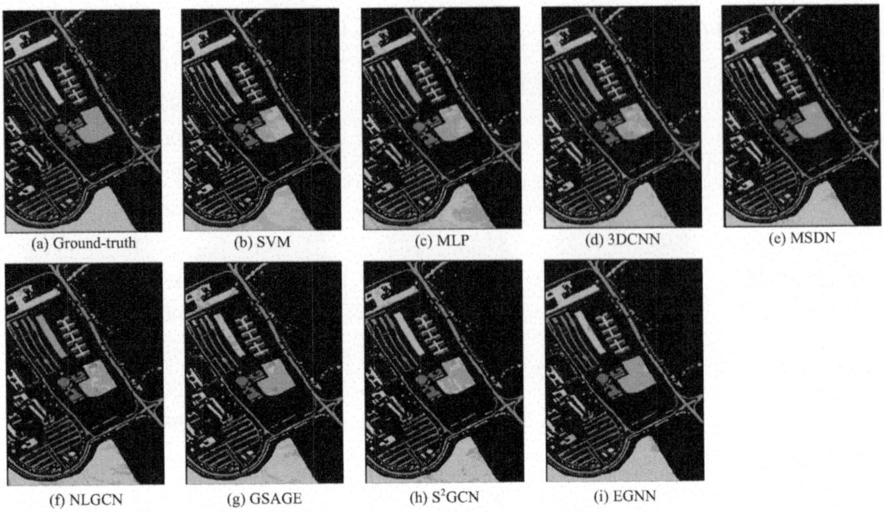

Fig. 5.4 Classification maps of comparison methods on PU dataset

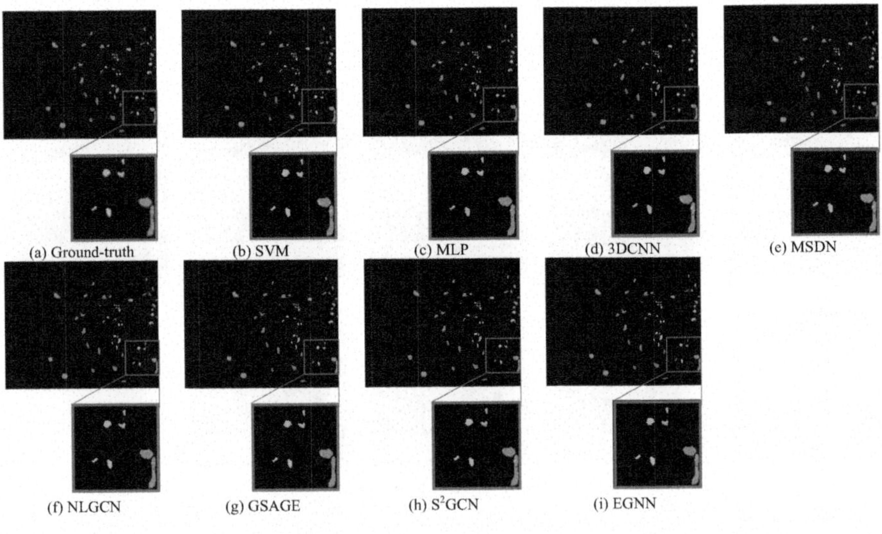

Fig. 5.5 Classification maps of comparison methods on KSC dataset

Table 5.4 Comparison of training time(s) of different methods on three datasets

Datasets	3DCNN	MSDN	NLGCN	GSAGE	S2GCN	EGNN
IP	248	1245	48.2	43.4	105	56.3
PU	456	1875	59.5	49.8	247	183
KSC	589	2122	75.8	63.4	387	282

Table 5.5 OA, AA and Kappa coefficients achieved by different convolution operation

Name	Indian Pines		Pavia University	
	GCN	EdgeConv	GCN	EdgeConv
OA	90.25	**93.12**	91.39	**95.31**
AA	93.88	**95.78**	92.10	**96.28**
Kappa	88.87	**92.14**	88.70	**93.84**

EGNN, its computational time consumption is still far lower than that of the general CNN-based methods. Taking into account both training efficiency and classification accuracy, the proposed EGNN method achieves a superior performance.

5.4.3 *Influence of EdgeConv*

To illustrate the effectiveness of edge convolution, we replace EdgeConv with traditional GCN. The results on two datasets are shown in Table 5.5, from which we can see that EdgeConv consistently yields better results than GCN, which demonstrates that EdgeConv can help improve the learning performance. This is because EdgeConv aggregates information according to the relationships between a point and its neighbors. Therefore, the EdgeConv can incorporate more discriminative information by alleviating the negative of samples outside the category.

5.4.4 *Impact of the Number of Labeled Examples*

In most cases, labeled pixels in HSI are usually rare and expensive to acquire. Thus, it is essential to evaluate the small sample learning capacity of an HSI classification model. We further conduct experiments to investigate the classification accuracies of all the compared methods under different numbers of labeled examples. The number of labeled examples per class ranges from 5 to 30. The results are shown in Fig. 5.6. We can observe that the performance of our method can be improved by increasing the number of labeled examples. Moreover, our method can achieve relatively good results with only 20 examples per class.

Fig. 5.6 OAs under different numbers of labeled examples per class on Pavia University

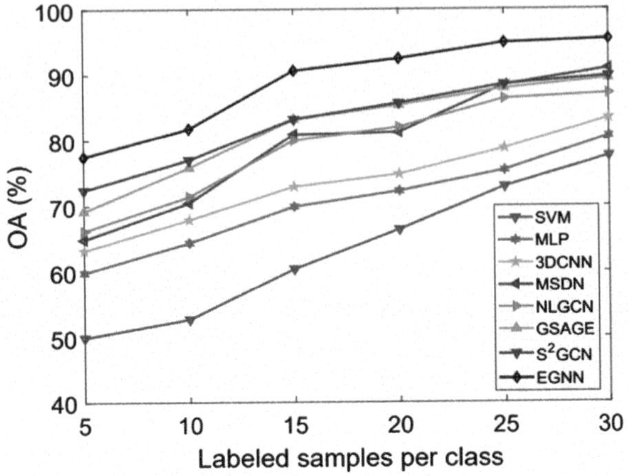

5.5 Conclusion

In this article, an EdgeConv-based GNN is proposed for hyperspectral image classification. We first partition the entire image into many spatially connected superpixels to reveal the strong spectral-spatial structure of HSI. By treating each superpixel as a graph node and establishing the adjacency relations between superpixels, we construct a graph to model the relationship between different regions in HSI, which is further fed into a novel end-to-end network architecture that combines GNN and CNN for superpixel and pixel-level feature fusion without loss of resolution. In the GNN, feature messages are passing through EdgeConv, which has shown its advantages over traditional GCN in the task of HSI classification. Experimental results reveal that our method significantly outperforms its counterparts in terms of OA, AA, and Kappa.

References

1. E. Blanzieri, F. Melgani, Nearest neighbor classification of remote sensing images with the maximal margin principle. IEEE Trans. Geosci. Remote Sens. **46**(6), 1804–1811 (2008)
2. Y. Tarabalka, M. Fauvel, J. Chanussot, J.A. Benediktsson, Svm- and mrf-based method for accurate classification of hyperspectral images. IEEE Geosci. Remote Sens. Lett. **7**(4), 736–740 (2010)
3. M. Fauvel, J.A. Benediktsson, J. Chanussot, J.R. Sveinsson, Spectral and spatial classification of hyperspectral data using svms and morphological profiles. IEEE Trans. Geosci. Remote Sens. **46**(11), 3804–3814 (2008)
4. X. Sun, Q. Qu, N.M. Nasrabadi, T.D. Tran, Structured priors for sparse-representation-based hyperspectral image classification. IEEE Geosci. Remote Sens. Lett. **11**(7), 1235–1239 (2014)
5. P. Yang, L. Tong, B. Qian, Z. Gao, J. Yu, C. Xiao, Hyperspectral image classification with spectral and spatial graph using inductive representation learning network. IEEE J. Select. Top. Appl. Earth Observ. Remote Sens. **14**, 791–800 (2021)

6. R. Hang, Z. Li, Q. Liu, P. Ghamisi, S.S. Bhattacharyya, "Hyperspectral image classification with attention-aided cnns. IEEE Trans. Geosci. Remote Sens. **59**(3), 2281–2293 (2020)
7. R. Hang, F. Zhou, Q. Liu, P. Ghamisi, Classification of hyperspectral images via multitask generative adversarial networks. IEEE Trans. Geosci. Remote Sens. **59**(2), 1424–1436 (2020)
8. M. Lan, Y. Zhang, L. Zhang, B. Du, Global context based automatic road segmentation via dilated convolutional neural network. Inform. Sci. **535**, 156–171 (2020)
9. Y. Li, H. Zhang, Q. Shen, Spectral–spatial classification of hyper-spectral imagery with 3D convolutional neural network. Remote Sens. **9**(1), 67 (2017)
10. S. Mei, J. Ji, Y. Geng, Z. Zhang, X. Li, Q. Du, Unsupervised spatial–spectral feature learning by 3d convolutional autoencoder for hyperspectral classification. IEEE Trans. Geosci. Remote Sens. **57**(9), 6808–6820 (2019)
11. X. Tang, F. Meng, X. Zhang, Y.-M. Cheung, J. Ma, F. Liu, L. Jiao, Hyperspectral image classification based on 3-d octave convolution with spatial–spectral attention network. IEEE Trans. Geosci. Remote Sens. **59**(3), 2430–2447 (2021)
12. R. Hang, Z. Li, P. Ghamisi, D. Hong, G. Xia, Q. Liu, Classification of hyperspectral and lidar data using coupled cnns. IEEE Trans. Geosci. Remote Sens. **58**(7), 4939–4950 (2020)
13. A. Qin, Z. Shang, J. Tian, Y. Wang, T. Zhang, Y.Y. Tang, Spectral spatial graph convolutional networks for semisupervised hyperspectral image classification. IEEE Geosci. Remote Sens. Lett. **16**(2), 241–245 (2019)
14. L. Mou, X. Lu, X. Li, X.X. Zhu, Nonlocal graph convolutional networks for hyperspectral image classification. IEEE Trans. Geosci. Remote Sens. **58**(12), 8246–8257 (2020)
15. D. Hong, L. Gao, J. Yao, B. Zhang, A. Plaza, J. Chanussot, Graph convolutional networks for hyperspectral image classification. IEEE Trans. Geosci. Remote Sens. **645**, 1–13 (2020)
16. S. Wan, C. Gong, P. Zhong, B. Du, L. Zhang, J. Yang, Multiscale dynamic graph convolutional network for hyperspectral image classification. IEEE Trans. Geosci. Remote Sens. **58**(5), 3162–3177 (2020)
17. A. Sha, B. Wang, X. Wu, L. Zhang, Semisupervised classification for hyperspectral images using graph attention networks. IEEE Geosci. Remote Sens. Lett. **18**(1), 157–161 (2021)
18. Y. Ding, X. Zhao, Z. Zhang, W. Cai, N. Yang, Graph sample and aggregate-attention network for hyperspectral image classification. IEEE Geosci. Remote Sens. Lett. **65**, 1–5 (2021)
19. Y. Ding, X. Zhao, Z. Zhang, W. Cai, N. Yang, Multiscale graph sample and aggregate network with context: aware learning for hyperspectral image classification. IEEE J. Select. Top. Appl. Earth Observ. Remote Sens. **14**, 4561–4572 (2021)
20. Y. Wang, et al. Dynamic graph cnn for learning on point clouds. ACM Transact. Gr. (tog) **38**(5), 1–12 (2019)
21. J. Gilmer, S.S. Schoenholz, P.F. Riley, O. Vinyals, G.E. Dahl, Neural message passing for quantum chemistry, in *International Conference on Machine Learning. PMLR* (2017), pp. 1263–1272
22. G. Li, M. Mueller, G. Qian, I.C. Delgadillo Perez, A. Abualshour, A.K. Thabet, B. Ghanem, Deepgcns: making gcns go as deep as cnns. IEEE Trans. Pattern Anal. Mach. Intell. **85**, 1 (2021)
23. C. Zhang, G. Li, S. Du, Multi-scale dense networks for hyperspectral remote sensing image classification. IEEE Trans. Geosci. Remote Sens. **57**(11), 9201–9222 (2019)

Chapter 6
Unifying Label Propagation and Graph Sparsification for Hyperspectral Image Classification

6.1 Introduction

Hyperspectral image (HSI) is a branch of optical remote sensing data, which has advantages of high spectral resolution and detailed spatial structure [1]. Among the applications in HSI processing, hyperspectral image classification (HSIC) is a fundamental yet challenging problem [2, 3], which aims to assign a specific label to each pixel in the image. HSIC has been widely applied to many scenarios, such as military target detection, vegetation monitoring, and disaster prevention and control.

Various approaches for HSIC have been presented during the last few decades. The early studies are mainly based on conventional pattern recognition methods, such as K-nearest neighbor classifier [4], support vector machines (SVM) [5, 6], and sparse representation [7]. The performance of these HSIC methods, on the other hand, is highly dependent on the handcrafted characteristics, which are heavily rely on professional experience and are quite empirical. In addition, with the development of the deep learning-based techniques, convolutional neural networks (CNN) has proven the effectiveness in image processing by automatically extracting discriminative features. Many variants of CNN-based models have been devised, ranging from 1D-CNN to 3D-CNN approaches [8] and from single networks to hybrid networks [8–10]. Although existing CNN-based approaches have shown considerable success, they require a large number of training parameters and are prone to overfitting due to a lack of training data. Meanwhile, CNN tends to blur the classification border by extracting features using a fixed shape kernel around the central pixel [11]. As a result, accurately classifying HSI remains a challenge.

To build more flexible models, many GCN-based HSIC methods have been proposed by learning features from the constructed graph structure. Generally, there can be two categories according to the types of adjacent graph [12, 13]. One is pixel-level adjacent graph. This type of graph can directly propagate information between nearby and distant regions, while it treats each pixel as a node in the graph, this

Y. Ding et al., *Graph Neural Network for Feature Extraction and Classification of Hyperspectral Remote Sensing Images*, Intelligent Perception and Information Processing, https://doi.org/10.1007/978-981-97-8009-9_6

results in a vast amount of computation and limits its applicability. The superpixel-level adjacent graph is another commonly used graph for HSI classification [14], which is usually accomplished by a superpixel segmentation algorithm. However, this graph only focuses on the local structure of the data. In order to capture the long-range contextual information, many works that unify GCN and CNN [15, 16], as well as multi-scale GCNs [17, 18] have been developed. However, multiple convolution layers are required to be stacked, which is highly inefficient and will bring potential concerns of over-smoothing.

Aiming to address the above issues of the superpixel-level adjacent graph, our key insight is to effectively encode the local spatial and global spectral information of HSI into the constructed graph. However, due to the limited training samples and complex spatial topologies in HSIs, it is difficult for a network to automatically learn the intricate relationships between distributed and difform land covers automatically [19]. As a result, we use the simple linear iterative clustering (SLIC) approach to divide the whole image into several spatially connected superpixels, which are then turned into a fully connected network based on their spatial proximity and spectral similarity. In order to avoid potentially misleading message propagation, we apply the neural sparsification technique to remove potentially task-irrelevant edges. Recent work in [20] has proven that increasing the accuracy of label propagation algorithm's predictions is equivalent to increasing the strength of edges between the nodes of the same class, which inspires us to set proper weights on the obtained graph edges for better classification results under the guidance of label propagation. Overall, the main contributions of this paper are summarized as follows:

- We incorporate both local and global information by a learnable graph, which will automatically decide whether the model will focus more on spatially similar nodes or on nodes with similar spectral characteristics;
- Our method applies a sparsification strategy that favors subsequent classification task by removing potentially task-irrelevant edges;
- Label propagation (LP) serves as regularization to assist the graph network in learning proper edge weights that benefit the separation of different node classes.

6.2 Preliminaries

The intuition of GCN lies on the smoothing assumption [21], i.e., nodes in the same class are more likely to be densely connected. The smoothing will make the features similar, which makes the following classification task much easier. However, it is hard to directly define a proper graph structure only by the given 3D structural data of HSI. Therefore, the main motivation of our paper is to provide a reliable graph structure for HSI with the help of label propagation algorithm (LPA) and graph neural sparsification. In this section, we recap the preliminaries of GCN and LPA.

6.2.1 Label Propagation Algorithm

Let $\mathcal{G} = (\mathcal{V}, \mathbf{A}, \mathbf{V}, \mathbf{Y})$ represents the given undirected graph, where $\mathcal{V} = \{v_1, v_2, \cdots, v_n\}$ is the set of nodes, $\mathbf{A} \in \mathbb{R}^{n\times n}$ is the adjacency matrix with a_{ij} (the ij-th entry of $\mathbf{A}$) defined as the weight of the edge connecting v_i and v_j, $\mathbf{V} \in \mathbb{R}^{n\times d}$ is the feature matrix and $\mathbf{Y} \in \{0,1\}^{n\times c}$ is labels of nodes, where c is the number of classes. LPA assumes that nearby nodes are more likely to have the same label, and it works by propagating labels iteratively along the edges [22, 23]. The formulation of LPA in iteration k can be defined as follows

$$\mathbf{Y}^{(k+1)} = \mathbf{D}^{-1}\mathbf{A}\mathbf{Y}^{(k)} \tag{6.1}$$

where $\mathbf{Y}^{(k)}$ is the soft label matrix in iteration k, $\mathbf{D}$ is the diagonal matrix for $\mathbf{A}$ with entries $d_{ii} = \sum_j a_{ij}$. Noting that $\mathbf{Y}^0$ is the initial label matrix $\mathbf{Y}$, which consists of one-hot label indicator vectors (labeled nodes) or zero vectors (unlabeled nodes) otherwise. After k-th iteration, the predicted label distribution for unlabeled nodes can aggregate the true label information from labeled nodes in their k-nearest neighbors.

6.2.2 Graph Convolutional Network

Graph convolutional network is a multi-layel neural network that propagates and transforms node features cross a given graph [24]. The output of layer k in GCN is

$$\mathbf{X}^{(k+1)} = \sigma\left(\widehat{\mathbf{A}}\mathbf{X}^{(k)}\mathbf{W}^{(k)}\right), \tag{6.2}$$

where σ denotes a non-linear activation function (typically ReLU). The normalized adjacency matrix is defined as $\widehat{\mathbf{A}} = \widetilde{\mathbf{D}}^{-1/2}(\mathbf{A}+\mathbf{I})\widetilde{\mathbf{D}}^{-1/2}$. Compared with the standard adjacency matrix $\mathbf{A}$, the self-loops are added, and the matrix elements are rescaled through the diagonal node degree matrix $\widetilde{\mathbf{D}}$. We can see that GCN calculates new features of a vertex as a weighted average of features of its neighbors on a graph, which allows the vertices in the same cluster to have similar features, making the subsequent classifications much easier.

6.3 Approach

The purpose of HSIC is to predict the labels of all the pixels in the HSI by a classification model given the supervised data set. As illustrated in Fig. 6.1, we present an LPA-enhanced Graph Neural Sparsification Network (LPAGSN) for the HSI classification model to adaptively capture both local and global contextual information.

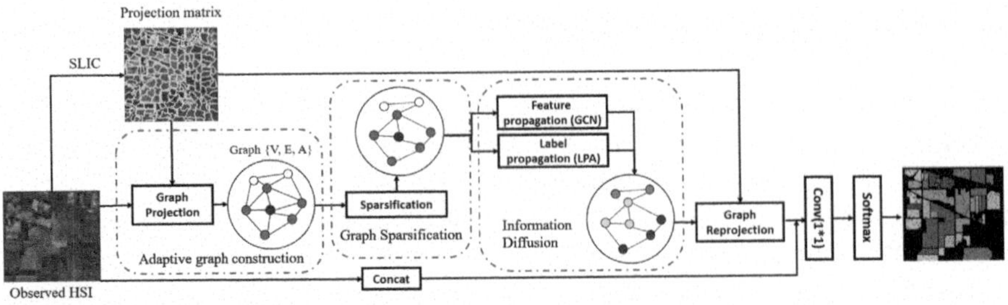

Fig. 6.1 Schematic depiction of the proposed LPA-enhanced Graph Neural Sparsification Network (LPAGSN). LPAGSN first assigns nearby pixels with similar features to the same vertex after graph projection. And then, a learned adjacent matrix that captures both local and global structure is obtained through adaptive graph construction. Under the guidance of sparsification, the graph removes most task-irrelevant edges and is subsequently fed into a two-layer GCN to get the high-level feature map. At the same time, LPA serves as regularization to assist the GCN in learning proper edge weights. After the graph reprojection, the corresponding coarse segmentation result is obtained. Finally, the refined feature map is fused with the original feature to get the final refined classification result

LPAGSN takes the original image data $\mathbf{X} \in \mathbb{R}^{H\times W\times B}$, the labels $\mathbf{Y} \in \{0,1\}^{H\times W\times C}$ of pixels in the training set, and the projection matrix $\mathbf{Q} \in \mathbb{R}^{M\times N}$ produced by a superpixel segmentation algorithm as input, where H, W, B, C, M, and N represent the height, width, number of bands, number of classes, number of pixels, and number of superpixels respectively. By treating superpixels as the vertexes in the graph, we use graph projection and reprojection to project features from image pixels to graph representation $\mathbf{V} \in \mathbb{R}^{N\times B}$ and transform graph features $\tilde{V} \in \mathbb{R}^{N\times B\prime}$ back to the image space. In the process of graph learning, a novel graph neural network that consists of three main modules is applied to realize feature and label propagation across a learnable graph. Each module will be described in depth in the following part.

Graph Projection and Reprojection. GCN-based models commonly work on superpixel-based nodes rather than pixel-based nodes to minimize computational complexity while preserving the local structure of HSI. Similarly, we split the whole image into multiple spatially related superpixels using the simple linear iterative clustering (SLIC) approach. Each superpixel is treated as a graph node, with the average spectral signatures of each superpixel serving as the node feature. The graph projection assigns $\mathbf{X}$ to a set of vertices and gets the feature matrix $\mathbf{V}$ of the vertices in the graph by the form of matrix multiplication as follows

$$\mathbf{V} = \widehat{\mathbf{Q}}^T \text{Flatten}(\mathbf{X}), \tag{6.3}$$

where Flatten $(\cdot)$ denotes flattening the HSI data by the spatial dimension, and $\widehat{\mathbf{Q}}$ is the normalized $\mathbf{Q}$ by column, i.e., $\widehat{\mathbf{Q}}_{i,j} = \mathbf{Q}_{i,j}/\sum_m \mathbf{Q}_{m,j}$. The projection matrix $\mathbf{Q} \in \mathbb{R}^{M\times N}$ introduced by SLIC is defined as

$$\boldsymbol{Q}_{i,j} = \begin{cases} 1, \text{ if } \mathbf{x}_i \in \mathcal{S}_j, \\ 0, \text{ otherwise}, \end{cases} \tag{6.4}$$

where $\mathcal{S}_i$ is the i-th superpixel that consists of several homogeneous pixels.

We perform graph reprojection to transform the resulting features back to the original coordinate space. The graph reprojection operation takes the inputs of transformed vertex features $\widetilde{\mathbf{V}}$ and the assignment matrix $\mathbf{Q}$, and produces the corresponding 2D feature map $\widetilde{\mathbf{X}}$. This operation is defined as

$$\widetilde{\mathbf{X}} = \text{Reshape}\left(\mathbf{Q}\,\widetilde{\mathbf{V}}\right), \tag{6.5}$$

where Reshape$(\cdot)$ denotes restoring the spatial dimension of the flattened data.

Adaptive Graph Construction. The graph (which is the adjacency matrix in our formulation) is constructed according to the similarity between different nodes. Denote $\mathbf{A}_{ij}$ as the (i, j)-element of the adjacency matrix $\mathbf{A}$, we have

$$\mathbf{A}_{ij} = \mathbf{w}^T\left[\mathbf{W}^T\mathbf{v}_i||\mathbf{W}^T\mathbf{v}_j\right] + b_{(\mathbf{v}_i,\mathbf{v}_j)}, \tag{6.6}$$

where $\mathbf{w}$ and $\mathbf{W}$ are weight parameters, $b_{(v_i,v_j)} = 1$ if the i-th region and the j-th region are spatially related, and otherwise $b_{(v_i,v_j)} = 0$. Compared to the adjacency graph described in [18], where the receptive field is restricted to the neighbors, Eq. (6.6) provides both global and local information, and the learnable graph will automatically decide whether the model will focus more on spatially similar nodes or on nodes with similar spectral characteristics.

Graph Sparsification. Since the graph constructed by Eq. (6.6) is fully connected, which is of high computaional cost in real applications, we focus on k-neighbor graph for graph neural sparsification. The k-neighbor graph is obtained by repeatedly sampling k edges for each node in the original graph. To make samples differentiable, we apply Gumbel-Softmax [25] to generate differentiable discrete samples. Firstly, softmax function is employed to compute the probability to sample the edge, which is

$$\pi_{v_i,v_j} = \frac{\exp\left(\mathbf{A}_{i,j}\right)}{\sum_{w\in\mathbb{N}_{v_i}}\exp\left(\mathbf{A}_{i,w}\right)} \tag{6.7}$$

Then we can generate differentiable samples using Gumble-Softmax as follows

$$x_{v_i,v_j} = \frac{\exp\left(\left(\log\left(\pi_{v_i,v_j}\right) + \varepsilon_v\right)/\tau\right)}{\sum_{w\in\mathbb{N}_{v_i}}\exp\left(\left(\log\left(\pi_{v_i,w}\right) + \varepsilon_w\right)/\tau\right)} \tag{6.8}$$

where x_{v_i,v_j} is a scalar that represents whether to sample the edge between v_i and v_j, $\epsilon_{v_j} = -\log(-\log(s))$ with s randomly drawn from Uniform (0, 1), and τ is a hyperparameter that controls the interpolation between discrete distribution and continuous categorical densities.

Information Diffusion. We perform two-layer GCN and LPA to propagate information on the graph according to Eqs. (6.1) and (6.2). The corresponding loss function consists of two terms as follows

$$\mathcal{L} = \underset{\mathbf{W},\mathbf{A}}{\operatorname{argmin}} J_{gcn}(\mathbf{W}, \mathbf{A}) + \lambda J_{lpa}(\mathbf{A}), \tag{6.9}$$

where J denotes the cross-entropy over all labeled pixels, λ is a balancing regularization parameter. Noting that the second term $J_{lpa}(\mathbf{A})$ is only related to $\mathbf{A}$, decreasing the loss of LPA is equivalent to increasing the strength of edges between the nodes of the same class.

6.4 Experiments

In this section, experiments are conducted to evaluate classification performance of the proposed LPAGSN on three public HSI datasets, including the Indian Pines (IP), the Pavia University (PU), and the Kennedy Space Center (KSC) datasets. The Indian Pines dataset is composed of 200 spectral bands with an image size of 145×145. There are 16 terrain classes for research and analysis. The Pavia University dataset contains 115 spectral bands with a size of 610×340. This dataset includes nine land-cover classes. For the Kennedy Space Center dataset, it contains 176 bands with a size of 512×614. 13 classes representing various land cover types are defined for the research.

6.4.1 Experimental Settings

To quantitatively evaluate performance of different HSIC methods, three widely used metrics, including overall accuracy (OA), average accuracy (AA), and kappa coefficient are adopted. We randomly choose 30 labeled pixels in each class for training in all datasets, and choose 15 labeled pixels if the corresponding class has fewer than 30 samples. In addition, the learning rate and the number of training epochs are set to 0.002 and 1000, respectively. Nesterov Adam algorithm is chosen to optimize networks. The regularization parameter λ and the number of superpixels N are set to 1 and 1000, respectively. The value of k for each k-neighbor graph is 10 on three HSI datasets. The experiments are implemented in PyTorch and run on a Windows 10 machine with 3.80 GHz i7-10700 K CPU, 32 GB main memory, and RTX 3090 GPU.

6.4.2 Classification Results

To verify the superiority of our method, we compare the proposed LPAGSN model with several state-of-the-art HSIC methods: SVM, Multilayer Perception (MLP), 3DCNN [8], MSDN [26], NLGCN [27], S^2GCN [12], MSAGE [18], and CEGCN [16]. All of the reported results are based on an average of 10 training sessions to circumvent the bias owing to random sampling, with the top results highlighted in bold. The overall classification performance of all the compared methods on IP, PU, and KSC datasets are shown in Tables 6.1, 6.2, and 6.3. As is shown in the results, GCN-based approaches, such as MSAGE, CEGCN, and the proposed LPAGSN, outperform classical machine learning and deep learning models (i.e., SVM, MLP, 3DCNN). This is due to GCN's ability to learn the relationships between various land coverings and model their spatial topologies on graphs. Moreover, our method exceeds other alternatives substantially. This observation reveals that compared to previous GCN-based HSIC methods, our data-driven LPAGSN can offer better results due to the cohabitation of local spatial information and global spectral information automatically.

Table 6.1 Quantitative experimental classification results on IP, the best performances and the second-best performances are marked in bold and italic

No	SVM	MLP	3DCNN	MSDN	NLGCN	S^2GCN	MSAGE	LPAGSN
1	93.75	87.50	**100.0**	*96.77*	**100.0**	**100.0**	**100.0**	**100.0**
2	66.17	58.94	79.54	82.68	75.39	74.95	**91.86**	*87.08*
3	64.88	63.88	69.63	59.87	76.88	*84.78*	76.33	**93.85**
4	82.61	86.96	*96.14*	93.79	94.20	93.07	**100.0**	**100.0**
5	86.98	84.11	90.29	89.83	91.83	*93.53*	86.91	**95.84**
6	88.14	87.14	85.87	94.93	90.57	96.26	*97.39*	**99.56**
7	**100.0**	*92.31*	**100.0**	**100.0**	**100.0**	**100.0**	**100.0**	**100.0**
8	92.19	92.19	95.98	*99.04*	**100.0**	99.55	**100.0**	**100.0**
9	**100.0**	**100.0**	**100.0**	**100.0**	**100.0**	**100.0**	**100.0**	**100.0**
10	70.49	70.49	82.91	81.47	92.78	**93.81**	71.78	*92.95*
11	54.27	57.77	65.20	76.33	73.81	79.71	**93.29**	*93.22*
12	56.84	60.57	77.62	86.30	83.84	93.73	**97.11**	*94.66*
13	96.57	96.57	*99.43*	**100.0**	*99.43*	98.82	**100.0**	**100.0**
14	81.13	83.00	90.61	93.20	91.50	95.12	**100.0**	*99.94*
15	63.76	71.07	74.72	81.90	88.48	87.46	**99.42**	*99.40*
16	*98.41*	**100.0**	*98.41*	**100.0**	**100.0**	**100.0**	**100.0**	**100.0**
OA (%)	69.72	70.07	79.19	83.14	83.83	87.35	*91.52*	**94.68**
AA (%)	81.01	80.78	87.88	89.76	91.17	93.17	*94.63*	**97.28**
Kappa	66.00	66.31	76.55	80.73	81.70	85.61	*90.27*	**93.90**

Table 6.2 Quantitative experimental classification results on PU, the best performances and the second-best performances are marked in bold and italic

No	SVM	MLP	3DCNN	MSDN	NLGCN	S^2GCN	MSAGE	LPAGSN
1	69.78	73.63	82.25	*91.52*	86.81	87.73	90.22	**98.23**
2	73.60	77.98	80.75	85.15	83.22	88.12	**99.66**	*97.11*
3	83.91	83.95	*94.83*	90.96	91.25	88.08	94.34	**99.23**
4	86.98	91.76	86.85	**97.62**	*96.11*	94.32	94.03	90.45
5	99.16	98.63	99.92	**100.0**	99.62	*99.77*	**100.0**	**100.0**
6	76.66	82.54	77.04	87.41	86.40	74.83	*96.39*	**100.0**
7	94.62	91.85	90.23	94.05	*96.69*	93.30	91.18	**100.0**
8	82.04	77.17	85.60	95.09	87.46	96.24	*97.55*	**98.65**
9	99.67	**100.0**	98.58	98.91	**100.0**	*99.89*	*99.89*	93.54
OA (%)	77.54	80.59	83.35	89.39	87.11	89.74	*96.72*	**97.65**
AA (%)	85.16	86.40	88.45	93.41	91.95	92.80	*95.92*	**97.61**
Kappa	71.49	75.25	78.59	86.21	83.35	87.25	*95.64*	**96.53**

Table 6.3 Quantitative experimental classification results on KSC, the best performances and the second-best performances are marked in bold and italic

No	SVM	MLP	3DCNN	MSDN	NLGCN	S^2GCN	MSAGE	LPAGSN
1	88.95	88.29	93.57	96.72	*97.13*	96.79	97.11	**100.0**
2	84.37	82.16	74.65	89.20	90.94	*91.92*	90.38	**100.0**
3	84.91	84.07	85.40	96.90	94.65	**97.92**	95.93	*97.35*
4	54.37	21.22	22.52	54.50	59.46	54.23	*66.36*	**96.85**
5	64.96	60.99	84.73	84.74	*85.65*	84.89	**86.51**	80.15
6	58.69	46.93	79.90	79.90	75.58	87.04	*88.66*	**98.99**
7	91.20	85.60	98.00	94.67	95.20	*98.27*	**100.0**	**100.0**
8	80.32	80.25	69.58	86.78	93.17	*97.33*	93.43	**99.00**
9	73.22	90.78	81.63	90.61	96.88	*97.14*	**100.0**	**100.0**
10	85.00	90.99	94.92	**99.47**	97.99	96.39	98.10	**100.0**
11	98.53	94.34	*99.74*	98.46	98.92	**100.00**	99.48	94.60
12	83.00	83.81	82.47	86.68	92.79	95.03	**97.86**	*96.19*
13	98.48	98.83	**100.0**	*99.44*	99.79	**100.0**	**100.0**	**100.0**
OA (%)	84.44	84.04	85.87	91.83	93.70	94.70	*95.65*	**98.26**
AA (%)	80.46	77.56	82.24	89.08	90.63	92.07	*93.37*	**97.16**
Kappa	82.62	82.17	84.22	90.87	92.95	94.08	*95.13*	**98.05**

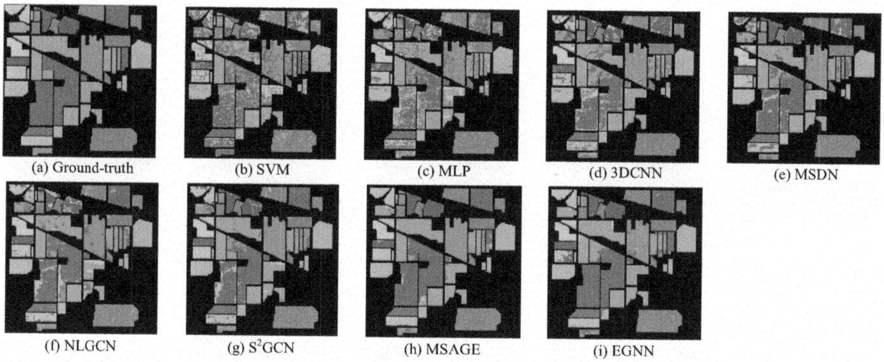

Fig. 6.2 Classification maps of comparison methods on IP dataset

For qualitative comparisons, the classification maps of three datasets are visualized in Figs. 6.2, 6.3 and 6.4 in comparison with the seven SOTA methods. From Figs. 5.2, 5.3, and 5.4, it can be clearly observed that the classification maps obtained by LPAGSN have the highest quality because they are closest to the ground-truth and have the least misclassification. Because only the spectral information is used without spatial features, the classification maps of SVM, MLP and NLGCN contain many scattered Salt-and-pepper noise. On the contrary, the classification map obtained by the spectral-spatial classifier is more compact and smoother. Specifically, due to the consideration of local spatial information and global spectral information, the classification map of our LPAGSN earns clearer boundary areas, verifying that the LPAGSN method can accurately extract discriminative features and help identify different features with similar characteristics.

6.4.3 Analysis of LPA

To illustrate the effectiveness of label propagation, we test the impact of λ on the performance of classification. Parameter λ represents the training weight of LPA loss term, and $\lambda = 0$ means training without the LPA loss term. The results on two datasets are shown in Fig. 6.5, where the value of λ is searched from $\{0,1, 2,5, 10,20,50\}$. From Fig. 5.5, we can see that the performance is boosted when the value of λ increases, then the accuracy stops increasing and decreases. This is because a suitable value of λ will help learn proper edge weights, while a large λ will replace the dominance of GCN and lead to a decrease in learning performance.

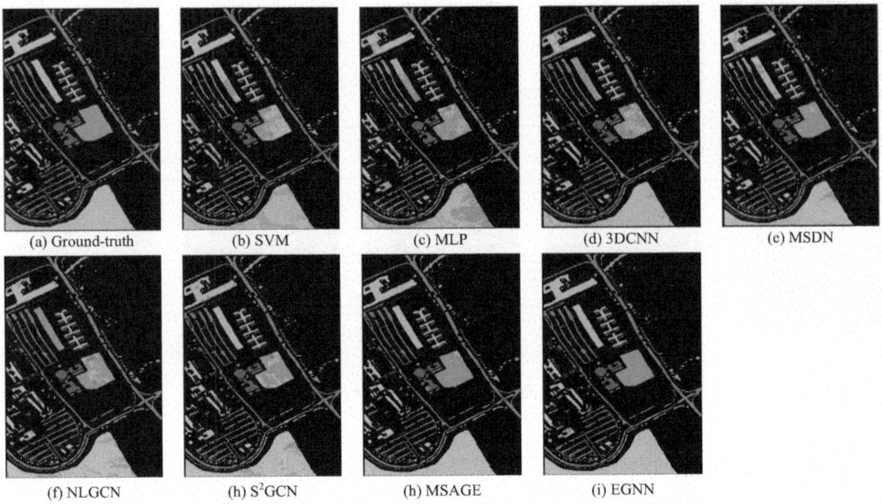

Fig. 6.3 Classification maps of comparison methods on PU dataset

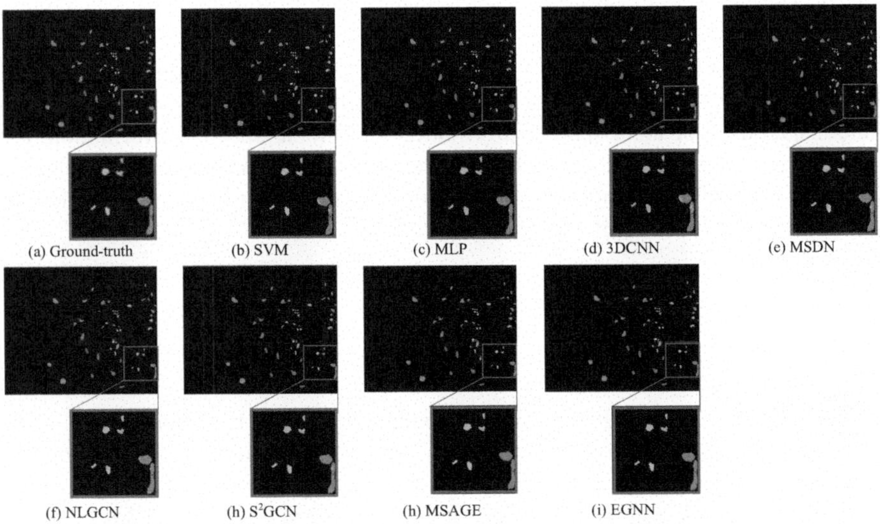

Fig. 6.4 Classification maps of comparison methods on KSC dataset

6.4.4 Analysis of the Graph Scale

There are two parameters that control the graph scale, i.e., the number of superpixels N and the number of neighbors k. N and k determine the size of the nodes in the graph and the sparsity of the graph, respectively. We conduct the parameter sensitivity experiments in terms of overall accuracy on two datasets. Experimental

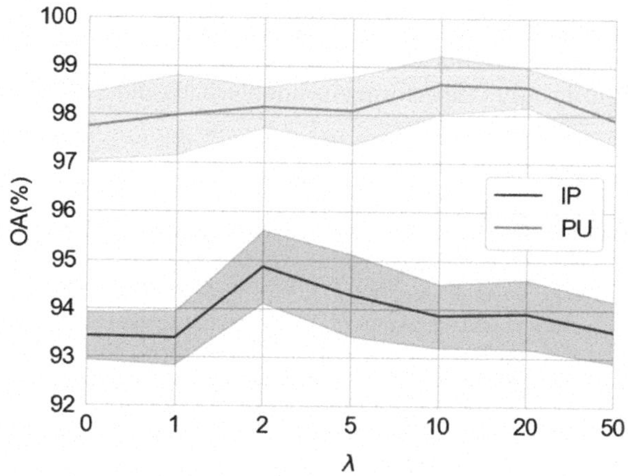

Fig. 6.5 Sensitivity to λ on IP and PU datasets

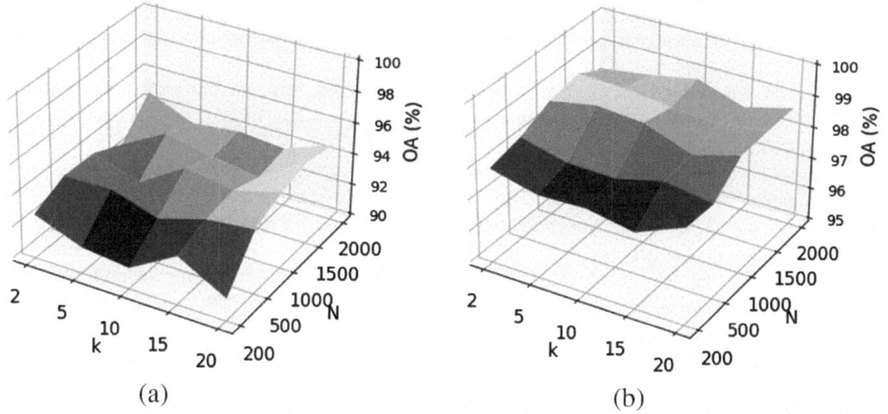

Fig. 6.6 Parameter sensitivity of k and N. **a** IP dataset. **b** PU dataset

results are shown in Fig. 6.6, where N is turned from {200,500,1000,1500,2000}, and k is searched from {2,5, 10,15,20}. We can conclude that our method is stable with varying k, while sensitive to N.

6.5 Conclusion

In this article, we propose a novel LPA-enhanced Graph Neural Sparsification Network (LPAGSN) for hyperspectral image classification. Different from the prior ways of constructing the adjacent graph, the proposed algorithm constructs a learnable graph that reveals the local and glocal structure of HSI. Besides, a differentiable sparsification technique is applied to favor subsequent classification task by removing potentially task-irrelevant edges. Label propagation is further performed to enhance the edge weights in the same class. Experimental results reveal that our method significantly outperforms its counterparts in terms of OA, AA, and Kappa.

References

1. L. Zhang, Q. Zhang, B. Du, X. Huang, Y.Y. Tang, D. Tao, Simultaneous spectral-spatial feature selection and extraction for hyperspectral images. IEEE Trans. Cybern. **48**(1), 16–28 (2016)
2. D. Hong, X. Wu, P. Ghamisi, J. Chanussot, N. Yokoya, X.X. Zhu, Invariant attribute profiles: a spatial-frequency joint feature extractor for hyperspectral image classification. IEEE Trans. Geosci. Remote Sens. **58**(6), 3791–3808 (2020)
3. J. Jiang, J. Ma, Z. Wang, C. Chen, X. Liu, Hyperspectral image classification in the presence of noisy labels. IEEE Trans. Geosci. Remote Sens. **57**(2), 851–865 (2019)
4. E. Blanzieri, F. Melgani, Nearest neighbor classification of remote sensing images with the maximal margin principle. IEEE Trans. Geosci. Remote Sens. **46**(6), 1804–1811 (2008)
5. Y. Tarabalka, M. Fauvel, J. Chanussot, J.A. Benediktsson, Svm- and mrf-based method for accurate classification of hyperspectral images. IEEE Geosci. Remote Sens. Lett. **7**(4), 736–740 (2010)
6. M. Fauvel, J.A. Benediktsson, J. Chanussot, J.R. Sveinsson, Spectral and spatial classification of hyperspectral data using svms and morphological profiles. IEEE Trans. Geosci. Remote Sens. **46**(11), 3804–3814 (2008)
7. X. Sun, Q. Qu, N.M. Nasrabadi, T.D. Tran, Structured priors for sparse-representation-based hyperspectral image classification. IEEE Geosci. Remote Sens. Lett. **11**(7), 1235–1239 (2014)
8. Y. Li, H. Zhang, Q. Shen, Spectral–spatial classification of hyper-spectral imagery with 3d convolutional neural network. Remote Sens. **9**(1), 67 (2017)
9. R. Hang, F. Zhou, Q. Liu, P. Ghamisi, Classification of hyper-spectral images via multitask generative adversarial networks". IEEE Trans. Geosci. Remote Sens. **59**(2), 1424–1436 (2021)
10. R. Hang, Z. Li, Q. Liu, P. Ghamisi, S.S. Bhattacharyya, "Hyperspectral image classification with attention-aided cnns. IEEE Trans. Geosci. Remote Sens. **59**(3), 2281–2293 (2021)
11. J. Bai, B. Ding, Z. Xiao, L. Jiao, H. Chen, A.C. Regan, Hyper-spectral image classification based on deep attention graph convolutional network. IEEE Trans. Geosci. Remote Sens. **54**, 1–16 (2021)
12. A. Qin, Z. Shang, J. Tian, Y. Wang, T. Zhang, Y.Y. Tang, Spectral spatial graph convolutional networks for semisupervised hyperspectral image classification. IEEE Geosci. Remote Sens. Lett. **16**(2), 241–245 (2019)
13. D. Hong, L. Gao, N. Yokoya, J. Yao, J. Chanussot, Q. Du, B. Zhang, More diverse means better: multimodal deep learning meets remote-sensing imagery classification. IEEE Trans. Geosci. Remote Sens. **59**(5), 4340–4354 (2021)
14. Y. Ding, X. Zhao, Z. Zhang, W. Cai, N. Yang, Y. Zhan, Semi-supervised locality preserving dense graph neural network with arma filters and context-aware learning for hyperspectral image classification. IEEE Trans. Geosci. Remote Sens. **32**, 1–12 (2021)
15. D. Hong, L. Gao, J. Yao, B. Zhang, A. Plaza, J. Chanussot, Graph convolutional networks for hyperspectral image classification. IEEE Trans. Geosci. Remote Sens. **59**(7), 5966–5978 (2021)
16. Q. Liu, L. Xiao, J. Yang, Z. Wei, "Cnn-enhanced graph convolutional network with pixel- and superpixel-level feature fusion for hyperspectral image classification. IEEE Trans. Geosci. Remote Sens. **59**(10), 8657–8671 (2021)
17. S. Wan, C. Gong, P. Zhong, B. Du, L. Zhang, J. Yang, Multiscale dynamic graph convolutional network for hyperspectral image classification. IEEE Trans. Geosci. Remote Sens. **58**(5), 3162–3177 (2020)

18. Y. Ding, X. Zhao, Z. Zhang, W. Cai, N. Yang, Multiscale graph sample and aggregate network with context-aware learning for hyperspectral image classification. IEEE J. Select. Top. Appl. Earth Observ. Remote Sens. **14**, 4561–4572 (2021)
19. Q. Liu, L. Xiao, J. Yang, Z. Wei, Multilevel superpixel structured graph u-nets for hyperspectral image classification. IEEE Trans. Geosci. Remote Sens. **65**, 1–15 (2021)
20. H. Wang, J. Leskovec, Unifying graph convolutional neural networks and label propagation (2020)
21. Q. Li, Z. Han, X.-M. Wu, Deeper insights into graph convolutional networks for semi-supervised learning, in *Thirty-Second AAAI Conference on Artificial Intelligence* (2018)
22. J. Jiang, J. Ma, X. Liu, Multilayer spectral–spatial graphs for label noisy robust hyperspectral image classification. IEEE Trans. Neural Netw. Learn. Syst. **33**(2), 839–852 (2022)
23. S. Jia, X. Deng, M. Xu, J. Zhou, X. Jia, Superpixel-level weighted label propagation for hyperspectral image classification. IEEE Trans. Geosci. Remote Sens. **58**(7), 5077–5091 (2020)
24. T.N. Kipf, M. Welling, Semi-supervised classification with graph convolutional networks (2016)
25. Zheng, Cheng, et al. Robust graph representation learning via neural sparsification. In *International Conference on Machine Learning*. PMLR (2020)
26. C. Zhang, G. Li, S. Du, Multi-scale dense networks for hyper-spectral remote sensing image classification. IEEE Trans. Geosci. Remote Sens. **57**(11), 9201–9222 (2019)
27. L. Mou, X. Lu, X. Li, X.X. Zhu, Nonlocal graph convolutional networks for hyperspectral image classification. IEEE Trans. Geosci. Remote Sens. **58**(12), 8246–8257 (2020)

Chapter 7
S^2GFormer: Exploring Relationship Between Transformer and Graph Convolution for Hyperspectral Image Classification

7.1 Introduction

Recently both graph convolutional neural networks (GCNs) and Transformers have shown promising progress in hyperspectral image (HSI) classification. Transformer-based methods have a great ability to model non-local interactions among spectral and spatial information, whereas GCNs tend to do well in exploiting neighborhood vertex interactions based on their unique aggregation mechanism.

In this chapter, we study how to comprehensively utilize the advantages of transformer and graph convolution by combing the two structures into a unified Transformer (Graphormer) to model both global and local interactions for HSI classification, and spatial-spectral features enhanced Graphormer framework (S^2GFormer) is proposed. Specifically, a Follow Patch mechanism is first proposed to transform the pixel in HSI to patches while preserving the local spatial features and reducing the computation cost. Moreover, a patch-wise spectral embedding block is designed to extract the spectral features of the patch, in which a neighborhood convolution is developed to extract comprehensive spectral information of HSI. Finally, a multi-layer Graphormer encoder module is proposed to extract the representative spatial-spectral features from the patch for HSI classification. In our network, the three aforementioned parts are jointly integrated into a unified end-to-end network, and each component benefits the other. Our method is evaluated using three widely used HSI data sets, and the experimental results demonstrate its suitability for HSI classification when compared with other state-of-the-art classifiers, particularly in scenarios with very limited labeled samples.

We can summarize the contributions as follows:

- A Follow Patch mechanism is designed to mine the relationship between HSI pixels while effectively preserving the local spatial features.
- A Patch-wise Spectral Embedding block is developed to extract comprehensive spectral information of HSI with a neighborhood convolution.

Y. Ding et al., *Graph Neural Network for Feature Extraction and Classification of Hyperspectral Remote Sensing Images*, Intelligent Perception and Information Processing, https://doi.org/10.1007/978-981-97-8009-9_7

- A multi-layer Graphormer encoder module is proposed to extract the representative spatial-spectral features from the patch for HSI classification. The Graphormer comprehensively utilizes the advantages of Transformer and graph convolution by combing the two structures into a unified Transformer to model both global and local interactions for HSI classification.
- We organize the remaining chapter as follows. In Sect. 7.2, some related works and concepts concerning the proposed method are reviewed. The details of the proposed S^2GFormer are introduced in Sect. 7.3. In Sect. 7.4, some experiments are carried out, and the experimental results are reported and analyzed. In Sect. 7.5, a comprehensive conclusion is drawn, and a brief outlook is discussed.

7.2 Vision Transformer

The Transformer architecture contains a sequence of Transformer layers. And each Transformer layer is composed of a self-attention module (shown as Fig. 7.1) and a position-wise feed-forward network. Given a self-attention module input $H = \left[h_1^T, h_2^T, \cdots, h_n^T\right]^T \in \mathbb{R}^{n\times d}$ where d and $h_i \in \mathbb{R}^{1\times d}$ denote the hidden dimension and the hidden representation at position i, respectively. The input H are mapped into three matrices Q (Query), K (Key), V (Value) by sharing the weights $W_q \in \mathbb{R}^{d\times d_k}$, $W_k \in \mathbb{R}^{d\times d_k}$, $W_v \in \mathbb{R}^{d\times d_v}$, respectively. Then the self-attention (SA) layer can be formed as

$$Q = HW_Q, K = HW_K, V = HW_V, \tag{7.1}$$

$$A = \frac{QK^{\mathrm{T}}}{\sqrt{d_K}}, \mathrm{Attn}(H) = \mathrm{softmax}(A)V \tag{7.2}$$

where A is the matrix that captures the similarity between Q and K.

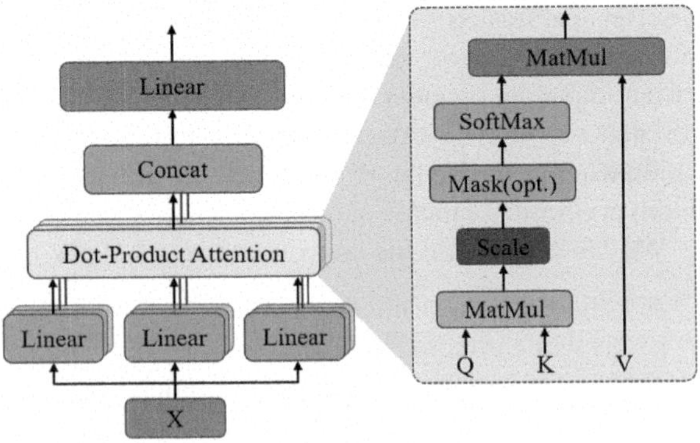

Fig. 7.1 Attention mechanism in transformer. Multi-head self-attention (left). Self-attention (right)

7.3 Proposed S^{2}GFormer for HIS Classification

In this section, we illustrate the architecture and each step of the proposed S^{2}GFormer in detail. Specifically, the overall framework description of the proposed method is first introduced to show the overall architecture (Sect. 7.3.1). Then, HSI pre-processing and Follow Patch mechanism are described (Sect. 7.3.2). Subsequently, the patch-wise spectral embedding block is introduced (Sect. 7.3.3). Furthermore, the multi-layer Graphormer Encoder is developed, including multi-head self-attention and graph residual block. Finally, the training and optimization details of the proposed method are detailed.

7.3.1 The Overall Framework Description

The proposed S^{2}GFormer mainly consists of three components, i.e., pixel-to-region transform module, patch-wise spectral embedding block, and multi-layer Graphormer Encoder module, and the overall framework of S^{2}GFormer is shown in Fig. 7.2. In the following, the three portions will be elaborated in detail.

- **HSI Pre-processing and Follow Patch**: The HSIs are transformed from pixels to regions while preserving the local spatial-spectral structure and reducing the computational complexity with Follow Patch mechanism.
- **Patch-wise Spectral Embedding Block**: The local patches are treated as the input. To adaptively extract the representative spectral features for patches while

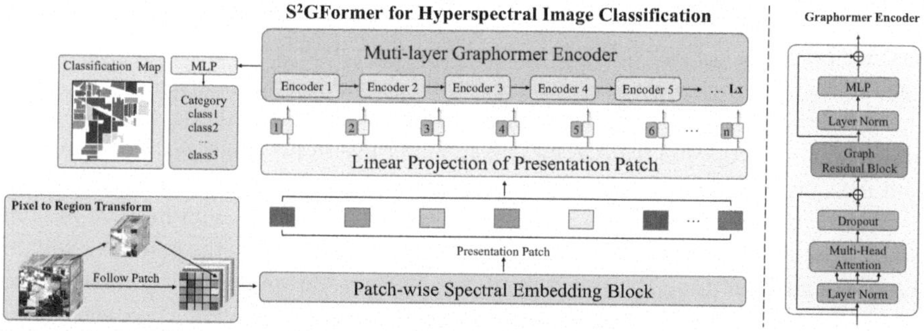

Fig. 7.2 Illustration of Graphormer with spatial-spectral features framework (S^{2}GFormer) for HSI classification. S^{2}GFormer mainly consists of three parts, i.e., pixel-to-region transform module, patch-wise spectral embedding block, and multi-layer Graphormer encoder module. The pixel-to-region transforms module is proposed to transform the pixels in HSI to locality-preserving regions using follow patch mechanism. The patch-wise spectral embedding block is developed to extract the spectral features of the patch. The multi-layer Graphormer encoder module is designed to extract the representative spatial-spectral features from the patch for HSI classification. Notably, the Graphomer encoder is newly designed, which is essentially different from Transformer

reducing the computational complexity, this module builds a neighborhood convolution mechanism by aggregating the local spectral information of each band in HSI.

- **Multi-layer Graphormer Encoder**: The spatial-spectral features are extracted for HSI classification. Furthermore, in Graphormer Encoder, both global and local spatial-spectral interactions are modeled by combining self-attentions and graph convolutions into a Transformer, newly designed and essentially different from the Transformer.

In our S^2GFormer, the HSI pre-processing and Follow Patch, patch-wise spectral embedding block, multi-layer Graphormer Encoder are joined in an end-to-end network for HSI classification, and each portion interacts with the other.

7.3.2 *HSI Pre-processing and Follow Patch*

HSI contains a large number of pixels, which is computationally intensive if we take each pixel as input for the S^2GFormer. To solve this problem, there are two main existing methods, i.e., the superpixel segmentation method [1] and the small sample method [2]. The superpixel segmentation method can segment the whole image into some superpixels, leading to reduce the number of input nodes. The small sample method intends to divide the graph into multiple subgraphs and compute the node features for each sample separately. However, the superpixel segmentation method inputs all the samples into the network at once, which places a great burden on memory. The small sample method is inefficient and cumbersome. Furthermore, the widely used patch methods cannot fully retain the local features of the image, which reduces the integrity of feature input. To overcome the aforementioned deficiencies, we proposed a Follow Patch mechanism, with which the input computational complexity of S^2GFormer can be reduced while preserving the local features of HSIs. Figure 7.3 shows the principal comparison of the three patch methods, which demonstrates that our proposed method can make better use of neighbor pixel information with the same patch size.

Given a classified pixel $x(x_{center}, y_{center})$, we take the patch with the size $m \times m$ around x as an input unit into the proposed S^2GFormer. Because the patch is a square area, we can determine the position of the patch in the image by calibrating the coordinates of the top left and bottom right positions of the patch. The coordinates of the upper-left corner coordinate of Follow Patch (x_p, y_p) can be expressed as

$$x_p = \begin{cases} x_{center} - \frac{m-1}{2}, & x_{center} > \frac{m-1}{2} \ and \ x_{center} < W - \frac{m-1}{2} \\ 0, & x_{center} \leq \frac{m-1}{2} \\ W - m - 1, & x_{center} \geq W - \frac{m-1}{2} \end{cases} \tag{7.3}$$

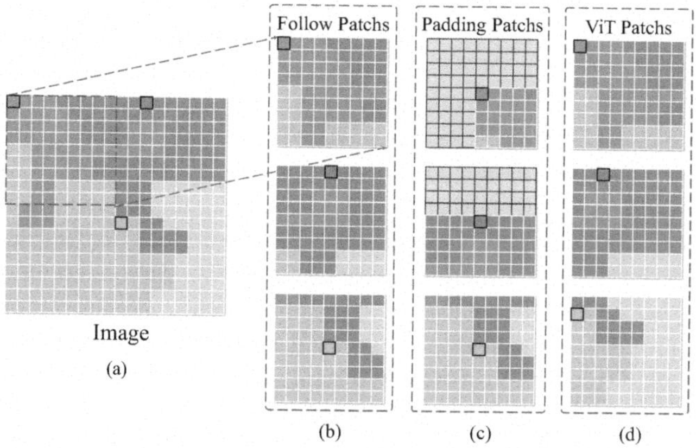

Fig. 7.3 Illustration of the follow patch mechanism. **a** Is the input image, **b**, **c**, and **d**, **b**, **c** and **d** are the results of follow patch method, padding patch method, and ViT patch method. Notably, the black bordered pixels in the figure are the target pixels

$$y_p = \begin{cases} y_{center} - \frac{m-1}{2}, & y_{center} > \frac{m-1}{2} \ and \ y_{center} < H - \frac{m-1}{2} \\ 0, & y_{center} \le \frac{m-1}{2} \\ H - m - 1, & y_{center} \ge H - \frac{m-1}{2} \end{cases} \tag{7.4}$$

where W and H are the width and height of the image, respectively. m is an odd number. Similarly, the lower-right corner coordinate (x_q, y_q) is formed as

$$x_q = \begin{cases} x_{center} + \frac{m-1}{2}, & x_{center} > \frac{m-1}{2} \ and \ x_{center} < W - \frac{m-1}{2} \\ \mathrm{m} - 1, & x_{center} \le \frac{m-1}{2} \\ W - 1, & x_{center} \ge W - \frac{m-1}{2} \end{cases} \tag{7.5}$$

$$y_q = \begin{cases} y_{center} + \frac{m-1}{2}, & y_{center} > \frac{m-1}{2} \ and \ y_{center} < H - \frac{m-1}{2} \\ \mathrm{m} - 1, & y_{center} \le \frac{m-1}{2} \\ H - 1, & y_{center} \ge H - \frac{m-1}{2} \end{cases} \tag{7.6}$$

With $((x_p, y_p), (x_q, y_q))$, we can accurately define a patch. From the above analysis, we can observe that the Follow Patch mechanism extends the spatial information of the image edge points and preserves richer spatial features, which is different from CNN selecting the convolution region in the form of Padding. In addition, the Follow Patch avoids the problem of classification pixel marginalization compared to the patch division in ViT. Compared with building global node information, the Follow Patch removes redundant spatial information in the form of aggregated local spatial information, making classification and computation more efficient.

7.3.3 Patch-Wise Spectral Embedding

We proposed a Follow Patch mechanism to preserve the local features of the HSI with patches. We still haven't solved the problem of extracting representative spectral features for the patch. The most mainstream method is to average the spectral values of all pixels in the patch [3], nevertheless, this method cannot solve the problem of spectral feature information redundancy, and is vulnerable to spectral noise interference. At present, spectral feature extraction pays aggressive attention to global spectral features, while ignoring local spectral features [4]. The Transformer can establish the nonlocal dependence of spectral sequence, but local information is ignored. To better fuse local information while adopting Transformer to establish global attention, a neighborhood convolution is developed, as shown in Fig. 7.4.

First, the features of the current and adjacent spectral bands are fused, we can form the neighbouring bands of the i-th spectrum X_i as

$$X^i_{group} = Concat(X_{\mathrm{m}}, X_{\mathrm{m+1}}, ..., X_i, ..., X_{m+\mathrm{k}-1}) \tag{7.7}$$

where k is the number of neighbouring spectrums, X^i_{group} is the neighbouring spectrum of the i-th spectrum, N is the total number of spectrums. The m is expressed as

$$\begin{cases} m = i - \frac{k-1}{2}, & i > \frac{k-1}{2} \ and \ i < N - \frac{k-1}{2} \\ m = 0, & i \leq \frac{k-1}{2} \\ m = N - k - 1, & i \geq N - \frac{k-1}{2} \end{cases} \tag{7.8}$$

After that, the original spectral information is combined with a residual cascade mechanism, which is defined as

$$X_i' = W_i X^i_{group} + b_i, i = 1, 2, ..., N \tag{7.9}$$

$$X_i'' = X + X_i' \tag{7.10}$$

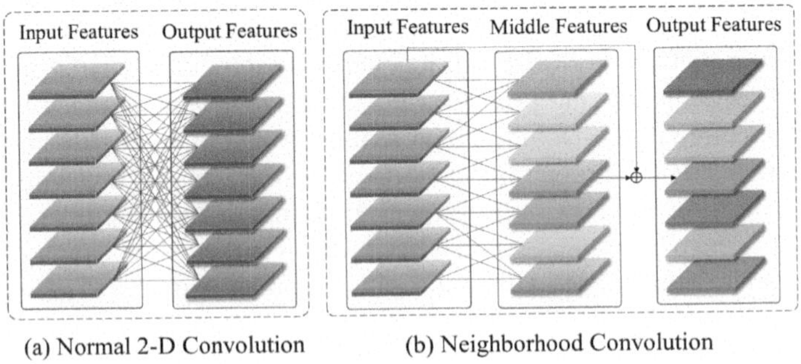

Fig. 7.4 Illustration of neighbourhood convolution

$$F = Concat(X_1'', X_2'', ..., X_N'') \tag{7.11}$$

where W_i and b_i are the weight and the bias of the i-th neighbourhood convolution to obtain the neighborhood feature X_i', the output feature X'' is obtained after residual concatenation, and the output feature map F is calculated with performing a cascade operation on X_i''.

To prevent overfitting, the BN and ReLU are utilized to regularize output to improve the classification performance, namely

$$\begin{aligned} F' &= \text{Re}LU(BN(F)) \\ &= \text{Re}LU\left(\gamma\frac{F-\mu}{\sigma}+\beta\right) \\ &= \max\left(0,\ \gamma\frac{F-\mu}{\sigma}+\beta\right) \end{aligned} \tag{7.12}$$

where F' is the output. μ and σ are the expectation and variance of a batch, respectively. γ and β are trainable parameters by network training.

7.3.4 Multi-layer Graphormer Encoder

In the S²GFormer framework, we proposed a Graphormer, combing the advantages of GCN and Transformer, to extract the spatial-spectral features of presentation patches. The architecture of our proposed Graphormer Encoder is demonstrated in Fig. 7.5. The Multi-layer Graphormer Encoder is comprised of four identical blocks. Each Graphormer Encoder contains six sub-modules, i.e., a Multi-Layer Perceptron (MLP), a Layer Norm, a Graph Residual Block, a Layer Dropout, a Multi-Head Self-Attention module, and a second Layer Norm. In our Graphormer Encoder, a graph convolution is embedded into the traditional Transformer encoder to model fine-grained local interactions, which is different from the traditional Transformer encoder.

(1) *Multi-Head Self-Attention:* In our Graphormer Encoder, a Multi-Head Self-Attention module is employed to learn contextual representation with several self-attention functions. Given an input patch sequence $\boldsymbol{P} = \{F_1', F_2', F_3' \ldots, F_n'\} \in \mathbb{R}^{n\times d}$, where d is the is the spectral dimension of patches. The input patch sequence is first projected to $\boldsymbol{Q}$, $\boldsymbol{K}$, and $\boldsymbol{V}$ by learning the trainable parameters $\{\boldsymbol{W}_Q, \boldsymbol{W}_K, \boldsymbol{W}_V\} \in \mathbb{R}^{d\times d}$, namely

$$\boldsymbol{Q}, \boldsymbol{K}, \boldsymbol{V} = \boldsymbol{P}\boldsymbol{W}_Q, \boldsymbol{P}\boldsymbol{W}_K, \boldsymbol{P}\boldsymbol{W}_V \in \mathbb{R}^{n\times d} \tag{7.13}$$

We can split three feature representations into h different subspaces, i.e., $\boldsymbol{Q} = [Q^1, Q^2, \ldots, Q^h]$, $Q^i \in \mathbb{R}^{n\times d}$, with which the self-attention can be performed for

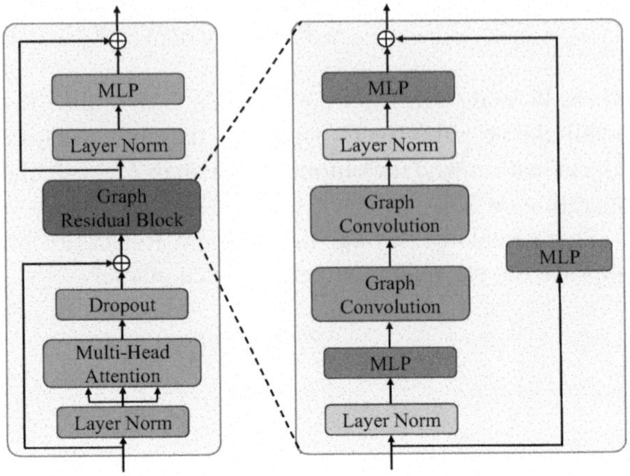

Fig. 7.5 Illustration of graphormer encoder

each subspace individually. Then the subspace output $\boldsymbol{Y}^h = \{y_1^h, y_2^h, \ldots, y_n^h\}$ can be calculated as

$$\boldsymbol{y}_i^h = Att\left(q_i^h, K^h\right)\boldsymbol{V}^h \in \mathbb{R}^{\frac{d}{h}} \tag{7.14}$$

where $Att(\cdot)$ is the attention function, with which the semantically relevant q_i^h to K^h is that quantified by scaling softmax and dot-product. Following that, we concatenate the output $\boldsymbol{Y}^h \in \mathbb{R}^{n\times\frac{d}{h}}$ to form the final output $\boldsymbol{Y}$.

(2) *Graph Residual Block:* Multi-Head Self-Attention is useful to extract long-range dependencies, however, the local information in complex patch sequence is easily ignored. Therefore, we propose a Graph Residual Block to refine the local spatial-spectral features contained in patch sequence by extracting the relationships between different patches.

In Eq. (7.14), the contextualized features $\boldsymbol{Y}$ generated by Multi-Head Self-Attention module, the local interactions can be improved by using a graph convolution, we can compute the output $\boldsymbol{Y}'$ as

$$\boldsymbol{Y}' = GraphConv\left(\overline{\boldsymbol{A}}, \boldsymbol{Y}; \boldsymbol{W}_G\right) = \sigma\left(\overline{\boldsymbol{A}}\boldsymbol{Y}\boldsymbol{W}_G\right) \tag{7.15}$$

where $\overline{\boldsymbol{A}} \in \mathbb{R}^{n\times n}$, $\boldsymbol{W}_G$ and $\sigma(\cdot)$ are adjacency matrix of a graph, trainable parameters, and activation function respectively. In our block, we adopt Gaussian Error Linear Unit (GeLU) to perform the network non-linearity according to BERT [5].

Following that, the output of Graphormer Encoder $\boldsymbol{Y}''$ can be expressed as

$$\boldsymbol{Y}'' = MLP\left(\sigma\left(\overline{\boldsymbol{A}}\boldsymbol{Y}'\boldsymbol{W}_G\right)\right) + MLP(\boldsymbol{Y}) \tag{7.16}$$

The graph residual block can explicitly encode the graph structure in the network, thus improving the spatial-spectral locality of the feature.

7.3.5 Training and Optimization Details

HSI preprocessing is demonstrated in Eqs. (7.3)–(7.6), the output is the input of the subsequent networks. Subsequently, a Patch-wise Spectral Embedding module is introduced to extract the spectral feature for patches. Finally, a Multi-layer Graphormer Encoder is designed to extract spatial-spectral features of patch sequence, and the output of S^{2}GFormer is $\boldsymbol{O}_G^{(final)}$. In our S^{2}GFormer, a cross-entropy error is utilized to penalize the difference between the label examples and the proposed Multi-layer Graphormer Encoder, which is written as

$$\boldsymbol{L} = -\sum_{z \in \mathbf{y}_G} \sum_{f=1}^{C} \boldsymbol{Y}_{zf} \, ln \boldsymbol{O}_{Gzf}^{(final)} \tag{7.17}$$

where $\mathbf{y}_G$, $\boldsymbol{Y}_{zf}$, and C are the labeled example set, label matrix, and number of classes. The parameters are updated by using Adam gradient descent similar to [6].

7.4 Experiments

In this section, extensive experiments and analyses are introduced to assess the HSI classification performance of the proposed S^{2}GFormer. Specifically, the three well-known HSI datasets and experimental setups are first introduced (Sect. 7.4.1); subsequently, the quantitative and visual HSI classification performances of S^{2}GFormer are compared with nine state-of-the-art algorithms (Sect. 7.4.2); moreover, the training process is visualized by embedding the t-SNE into the proposed method (Sect. 7.4.3), furthermore, the influence of label ratio and hyperparameters are analyzed (Sects. 7.4.4 and 7.4.5); finally, we implement some ablation experiments to evaluate the rationality of the proposed S^{2}GFormer design (Sect. 7.4.6).

7.4.1 Benchmark Datasets and Setups

In our experiments, three well-known HSI datasets were adopted, including the Indian Pines dataset (IP), the Salinas dataset, and the Houston 2013 (UH2013) dataset. The first IP dataset, one of the earliest used datasets, was acquired over Northwestern Indiana (USA) by the Airborne Visible/Infrared Imaging Spectrometer (AVIRIS) sensor. The HSI contains 145×145 pixels, and the spatial resolution is 20 m/pixel. It includes 200 continuous bands after discarding 20 noisy and water-absorbed bands ranging from 0.4 to 2.5 μm with 16 landcover classes. The second Salinas dataset is also collected by the AVIRIS sensors. It contains 512×217 pixels, and the spatial resolution is 3.7 m/pixel. It includes 204 continuous bands after removing noisy

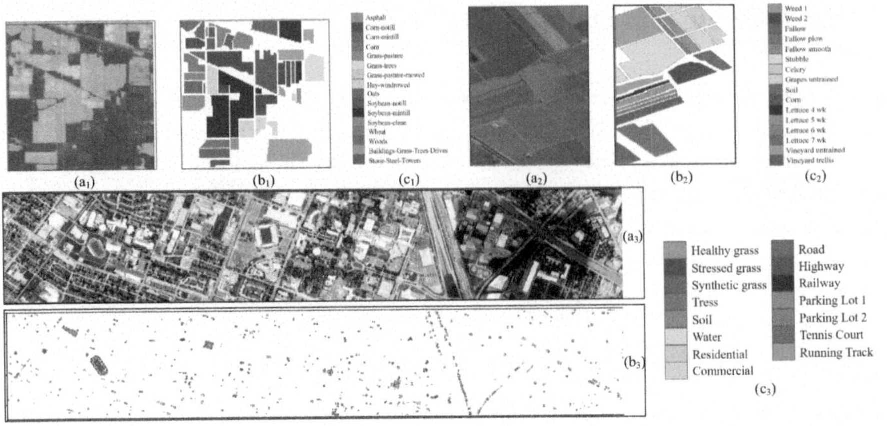

Fig. 7.6 Three dataset details. $\mathbf{a}_1$, $\mathbf{a}_2$, and $\mathbf{a}_3$ are the false-color maps of IP, Salinas, UH2013 datasets. $\mathbf{b}_1$, $\mathbf{b}_2$, and $\mathbf{b}_3$ are the ground-truth maps. c1, c2, and c3 are corresponding reference colors

and water-absorbed bands with 16 landcover categories. The third UH2013 dataset includes 349 $\times$ 1905 pixels with 15 landcover categories. Furthermore, it contains 144 continuous bands, ranging from 364 to 1046 nm. We show the true-color map, ground-truth map, and reference color of three datasets in Fig. 7.6. The details of three datasets of each class used in the evaluation experiment are demonstrated in Table 7.1.

We adopted four commonly used indices, including per-class accuracy, overall accuracy (OA), average accuracy (AA), and kappa coefficient (κ). In addition, we visualize the HSI classification maps obtained from all investigated models for qualitative comparison.

We select several recent representative baselines for the comparison experiments, including SVM-RBF, attention-based bidirectional long and short-term memory network (AB-LSTM), 2D CNN [7], convolutional RNN (CRNN) [8], miniGCN [2], spectral-spatial graph convolutional network (S^{2}GCN) [9], the Spectralformer (SF) [10], and SSFTT [11]. In these methods, SVM-RBF and AB-LSTM are traditional classification methods; 2D CNN and CRNN are CNN-based methods; miniGCN and S^{2}GCN are GCN-based methods; SF and SSFTT are state-of-the-art (SOTA) Transformer-based methods designed for HSI classification. In our S^{2}GFormer, five main parameters, i.e., epoch number T, patch size P_n, learning rate L, depth number V_d, and heads number V_h should be preset. Their optimal values are shown in Table 7.2 and will be analyzed in Sect. 7.4.4.

Furthermore, all investigated experiments are conducted on NVIDIA Titan RTX using the Pytorch framework. For a fair comparison, all methods are executed ten times for the sake of eliminating any bias caused by the random selection of training samples.

Table 7.1 The details of each land-cover class of three datasets used for evaluation in the experiments

No	Indian pines			Salinas			UH2013		
	Category name	Training	Testing	Category name	Training	Testing	Category name	Training	Testing
1	Asphalt	15	31	Weed 1	30	1979	Healthy-grass	30	1221
2	Corn-notill	30	1398	Weed 2	30	3696	Stressed-grass	30	1224
3	Corn-mintill	30	800	Fallow	30	1946	Synthetic-grass	30	667
4	Corn	30	202	Fallow-plow	30	1364	Tress	30	1214
5	Grass-pasture	30	453	Fallow-smooth	30	2648	Soil	30	1212
6	Grass-trees	30	700	Stubble	30	3929	Water	30	295
7	Grass-pasture-mowed	15	13	Celery	30	3549	Residential	30	1238
8	Hay-windrowed	30	448	Grapes untrained	30	11,241	Commercial	30	1214
9	Oats	15	5	Soil	30	6137	Road	30	1222
10	Soybean-notill	30	942	Corn	30	3248	Highway	30	1197
11	Soybean-mintill	30	2425	Lettuce-4 week	30	1038	Railway	30	1205
12	Soybean-clean	30	563	Lettuce-5 week	30	1897	Parking-Lot 1	30	1203
13	Wheat	30	175	Lettuce-6 week	30	886	Parking-Lot 2	30	439
14	Woods	30	1235	Lettuce-7 week	30	1040	Tennis-court	30	398
15	Buildings-grass-trees-drives	30	356	Vineyard-untrained	30	7238	Running-track	30	630
16	Stone-steel-towers	30	63	Vineyard-trellis	30	1777			

Table 7.2 Preset optimal ramenters in S^{2}GFormer

Dataset	T	P_n	L	V_d	V_h
IP	200	9×9	5×10^{-4}	3	4
Salinas	200	9×9	5×10^{-4}	3	4
UH 2013	200	9×9	5×10^{-4}	3	4

7.4.2 Comparison with SOAT Methods

In this section, following the experimental setups in Sect. 7.4.1, we perform HSI classification on the IP, Salinas, and UH2013 datasets and report the mean and the standard deviation of each quantitative metric (i.e., per-class accuracy, OA, AA, and κ) in detail in Tables 7.3, 7.4, and 7.5, in which the best performances and the second-best performances are marked in bold and underlined, respectively.

Classification results on IP dataset: The first experiment is conducted on the IP dataset to assess the classification performance of our S^{2}GFormer for small-scale HSI datasets with severe noise interference. The quantitative and corresponding qualitative classification results are summarized in Table 7.3 and Fig. 7.7, respectively.

From Table 7.3, due to extreme sample imbalance between different classes and severe noise interference, we can note that all other considered models perform poorly on IP image, however, the proposed S^{2}GFormer achieves 93.55, 96.56, and 92.62% in terms of OA, AA, and κ, which produces an improvement of 10.59, 5.01, and 11.91% compared with the second-best performances achieved by SSFTT.

Moreover, SSFTT and S^{2}GCN achieve relatively good classification results, which shows that the two methods have good adaptabilities to noise interference and class imbalance sample classification. However, miniGCN and SF have no advantages over other classification methods, owing to that the two algorithms are not specifically designed for HSI classification. Due to lacking the ability to extract high-lever features of HSI, the conventional methods (i.e., AB-LSTM and SVM-RBF) undoubtedly achieve poor results. Specifically, the OAs of the two methods are 56.68 and 66.96%, respectively. In addition, although the miniGCN and S^{2}GCN can dynamically update the graphs and fuse multiscale information for HSI classification, they cannot fully explore the relationship between spectral bands, resulting in unsatisfactory classification performance. Furthermore, compared with SSFTT, a patch-wise spectral embedding block is proposed in S^{2}GFormer to extract the presentive spectral features for HSI classification, enhancing the anti-interference of the algorithm. Hence, S^{2}GFormer achieves the smallest among all methods in the OA, AA, and κ values, indicating the best stability. In terms of the qualitative visual classification maps in Fig. 7.7, S^{2}GFormer obtains smoother and fewer misclassification map compared with other competitors. For example, in Soybean-mintill and Hay-windrowed, AB-LSTM and SVM-RBF contain lots of "salt and pepper" noises. We can note that SSFTT obtains relatively good classification results, nevertheless, it

Table 7.3 Quantitative experimental classification results on IP, the best performances and the second-best performances are marked in bold and italic

No	AB-LSTM	SVM-RBF	2D CNN	CRNN	miniGCN	S^2GCN	SF	SSFTT	S^2GFormer
1	41.94 ± 34.24	31.36 ± 3.65	*87.01 ± 0.00*	75.36 ± 0.26	68.32 ± 5.39	76.83 ± 3.97	47.31 ± 13.52	**100.00 ± 0.00**	**100.00 ± 0.00**
2	41.20 ± 15.40	58.16 ± 4.99	41.18 ± 6.81	38.35 ± 5.97	48.17 ± 7.31	57.29 ± 6.29	50.74 ± 5.01	*77.23 ± 2.92*	**85.77 ± 1.54**
3	12.62 ± 15.86	51.12 ± 5.52	46.92 ± 8.72	51.23 ± 8.01	46.28 ± 7.05	61.02 ± 5.37	57.04 ± 6.71	*77.46 ± 10.09*	**95.62 ± 0.28**
4	41.55 ± 27.07	30.90 ± 1.75	88.89 ± 2.40	79.66 ± 3.87	82.03 ± 4.24	78.66 ± 1.82	84.06 ± 6.35	**98.39 ± 1.39**	*98.07 ± 1.49*
5	76.97 ± 2.10	83.18 ± 0.69	73.36 ± 6.93	70.62 ± 4.01	76.02 ± 2.19	80.60 ± 1.85	72.33 ± 6.60	*93.75 ± 4.06*	**95.36 ± 0.18**
6	70.00 ± 23.64	91.13 ± 1.14	88.29 ± 4.15	85.27 ± 2.64	89.19 ± 3.09	92.28 ± 2.59	82.29 ± 0.91	*96.67 ± 1.36*	**98.43 ± 0.34**
7	43.59 ± 32.23	44.18 ± 10.37	**100.00 ± 0.00**	**100.00 ± 0.00**	95.13 ± 1.42	*98.33 ± 1.92*	**100.00 ± 0.00**	**100.00 ± 0.00**	**100.00 ± 0.00**
8	93.01 ± 4.86	95.79 ± 1.35	97.02 ± 1.00	*98.66 ± 1.28*	87.06 ± 1.57	91.36 ± 0.61	96.35 ± 1.24	**99.78 ± 0.32**	**99.78 ± 0.12**
9	60.00 ± 43.20	18.17 ± 1.37	**100.00 ± 0.00**	**100.00 ± 0.00**	*93.00 ± 1.97*	**100.00 ± 0.00**	**100.00 ± 0.00**	**100.00 ± 0.00**	**100.00 ± 0.00**
10	51.20 ± 18.09	56.76 ± 2.69	57.93 ± 3.73	58.03 ± 5.59	62.82 ± 4.33	68.26 ± 3.82	71.09 ± 4.68	*87.30 ± 2.73*	**91.08 ± 1.71**
11	52.04 ± 14.98	*73.59 ± 3.41*	59.64 ± 3.73	43.92 ± 5.27	66.08 ± 2.62	71.08 ± 2.01	53.46 ± 9.35	67.68 ± 7.54	**90.93 ± 1.37**
12	41.44 ± 2.36	46.22 ± 5.87	62.94 ± 5.66	58.30 ± 4.98	51.07 ± 5.88	62.35 ± 3.21	43.64 ± 4.48	*83.37 ± 2.76*	**90.94 ± 0.52**
13	97.71 ± 0.47	86.97 ± 0.51	99.05 ± 0.27	96.07 ± 0.95	89.31 ± 0.32	91.06 ± 1.00	97.71 ± 1.62	*99.62 ± 0.54*	**100.00 ± 0.00**
14	90.58 ± 3.19	94.43 ± 0.82	83.13 ± 0.60	81.28 ± 1.11	86.07 ± 0.92	91.06 ± 1.38	76.60 ± 5.15	*91.77 ± 1.61*	**99.60 ± 0.36**
15	54.21 ± 11.81	51.70 ± 2.47	80.43 ± 6.91	76.54 ± 5.90	82.39 ± 1.58	87.37 ± 1.93	85.49 ± 3.32	95.51 ± 3.58	**99.44 ± 0.75**
16	98.41 ± 2.24	83.17 ± 8.52	**100.00 ± 0.00**	96.01 ± 1.62	76.92 ± 1.25	85.93 ± 1.72	*98.94 ± 1.50*	96.30 ± 5.24	**100.00 ± 0.00**
OA (%)	56.68 ± 3.48	66.96 ± 2.25	65.83 ± 1.78	66.28 ± 0.99	68.73 ± 2.61	75.24 ± 1.97	65.24 ± 2.09	*82.96 ± 1.21*	**93.55 ± 0.47**
AA (%)	60.41 ± 2.84	62.30 ± 1.66	79.12 ± 1.21	75.58 ± 1.92	74.99 ± 0.41	80.84 ± 0.81	76.07 ± 0.95	*91.55 ± 0.97*	**96.56 ± 0.35**
Kappa	51.02 ± 3.55	62.81 ± 2.42	61.38 ± 1.99	68.21 ± 0.48	68.36 ± 1.94	74.71 ± 2.38	61.06 ± 2.10	*80.71 ± 1.29*	**92.62 ± 0.53**

Table 7.4 Quantitative experimental classification results on Salinas, the best performances and the second-best performances are marked in bold and italic

No	AB-LSTM	SVM-RBF	2D CNN	CRNN	miniGCN	S²GCN	SF	SSFTT	S²GFormer
1	98.50 ± 1.33	98.70 ± 1.05	98.77 ± 0.80	99.34 ± 0.53	95.20 ± 0.82	*99.01 ± 0.44*	94.74 ± 0.93	**100.00 ± 0.00**	**100.00 ± 0.00**
2	97.06 ± 3.51	99.20 ± 0.55	98.70 ± 0.71	99.17 ± 0.21	98.61 ± 0.59	99.18 ± 0.59	98.82 ± 0.49	*99.49 ± 0.46*	**99.86 ± 0.13**
3	73.14 ± 16.50	92.85 ± 0.61	90.00 ± 1.19	96.54 ± 1.39	94.15 ± 0.94	*97.15 ± 2.76*	92.17 ± 1.99	90.10 ± 7.84	**99.98 ± 0.03**
4	*99.49 ± 0.27*	97.54 ± 0.17	98.75 ± 0.36	97.32 ± 0.27	93.07 ± 0.82	99.11 ± 0.55	96.26 ± 0.94	**99.76 ± 0.30**	99.29 ± 0.04
5	93.76 ± 2.99	*97.65 ± 0.65*	95.20 ± 1.07	**98.75 ± 0.89**	87.23 ± 1.14	97.55 ± 2.35	89.41 ± 1.83	97.13 ± 2.30	96.11 ± 1.35
6	96.77 ± 2.37	**99.89 ± 0.02**	99.40 ± 0.39	99.19 ± 0.77	95.13 ± 0.86	99.32 ± 0.35	99.20 ± 0.72	*99.71 ± 0.31*	99.50 ± 0.21
7	98.96 ± 0.38	98.76 ± 0.43	**99.38 ± 0.37**	98.67 ± 1.30	90.28 ± 0.62	90.06 ± 0.27	97.91 ± 1.99	99.03 ± 0.50	*99.30 ± 0.49*
8	41.24 ± 30.46	77.68 ± 1.55	72.55 ± 6.82	72.38 ± 3.98	69.31 ± 0.97	70.68 ± 5.20	73.07 ± 1.12	*73.95 ± 3.67*	**92.59 ± 1.02**
9	91.23 ± 6.09	*99.18 ± 0.24*	93.30 ± 0.93	97.29 ± 0.61	96.00 ± 1.40	98.32 ± 1.79	93.77 ± 1.42	98.59 ± 1.35	**99.62 ± 0.37**
10	47.84 ± 23.55	83.99 ± 1.33	88.24 ± 0.95	91.44 ± 1.64	88.29 ± 2.69	90.97 ± 2.59	91.39 ± 1.83	*94.28 ± 1.14*	**95.83 ± 0.43**
11	82.69 ± 11.92	89.42 ± 0.31	93.96 ± 2.26	96.82 ± 0.78	96.85 ± 0.26	98.00 ± 1.65	92.36 ± 3.31	*98.65 ± 0.27*	**99.79 ± 0.12**
12	95.01 ± 1.29	95.31 ± 0.35	99.33 ± 0.22	99.21 ± 0.32	96.73 ± 0.71	99.56 ± 0.59	99.05 ± 0.58	**99.95 ± 0.04**	*99.66 ± 0.21*
13	86.61 ± 15.19	94.17 ± 1.21	99.06 ± 0.93	97.29 ± 0.86	95.29 ± 1.31	97.83 ± 0.72	**99.40 ± 0.43**	97.86 ± 1.28	*99.32 ± 0.65*
14	92.98 ± 2.45	90.11 ± 4.68	96.73 ± 0.95	95.10 ± 1.73	93.66 ± 1.27	95.75 ± 1.65	96.60 ± 0.95	*98.91 ± 0.16*	**99.42 ± 0.63**
15	58.63 ± 30.10	60.59 ± 1.96	71.35 ± 3.32	76.33 ± 4.62	71.49 ± 2.61	70.36 ± 3.62	*81.58 ± 2.20*	81.10 ± 3.63	**93.27 ± 0.16**
16	88.37 ± 8.03	*97.04 ± 1.20*	92.61 ± 1.97	97.99 ± 0.61	95.08 ± 1.25	96.90 ± 1.97	93.72 ± 2.04	96.96 ± 2.05	**98.26 ± 0.60**
OA (%)	74.83 ± 1.60	86.31 ± 0.53	87.59 ± 1.49	87.64 ± 0.83	86.75 ± 1.97	88.39 ± 1.01	88.80 ± 0.21	*90.67 ± 1.52*	**96.92 ± 0.36**
AA (%)	83.89 ± 3.61	92.01 ± 0.15	92.96 ± 0.56	94.55 ± 0.57	91.02 ± 0.82	94.30 ± 0.47	93.09 ± 0.19	*95.34 ± 1.11*	**98.23 ± 0.07**
Kappa	72.21 ± 1.69	85.90 ± 0.58	86.21 ± 1.64	86.72 ± 1.01	83.29 ± 1.48	87.10 ± 1.12	87.57 ± 0.24	*89.64 ± 1.68*	**95.62 ± 0.31**

Table 7.5 Quantitative experimental classification results on uh2013, the best performances and the second-best performances are marked in bold and italic

No	AB-LSTM	SVM-RBF	2D CNN	CRNN	miniGCN	S^2GCN	SF	SSFTT	S^2GFormer
1	82.76 ± 3.29	86.98 ± 2.74	85.53 ± 3.42	82.45 ± 2.19	85.16 ± 3.72	96.30 ± 3.07	86.66 ± 2.51	*94.31 ± 1.39*	**98.24 ± 0.62**
2	80.31 ± 3.66	78.11 ± 3.62	81.02 ± 4.09	84.12 ± 3.64	75.16 ± 4.26	**98.57 ± 1.47**	91.34 ± 1.32	87.32 ± 2.61	*97.67 ± 0.62*
3	77.26 ± 3.82	80.58 ± 4.92	89.27 ± 1.98	91.56 ± 1.07	85.72 ± 1.35	*98.88 ± 0.43*	93.76 ± 0.79	92.36 ± 1.27	**100.00 ± 0.00**
4	76.09 ± 5.85	78.11 ± 6.31	86.31 ± 2.92	91.29 ± 3.78	89.32 ± 2.92	*97.68 ± 2.89*	90.31 ± 0.39	88.42 ± 0.92	**100.00 ± 0.00**
5	82.33 ± 3.92	87.36 ± 5.14	94.72 ± 1.51	*98.81 ± 0.62*	95.37 ± 1.28	97.66 ± 1.12	96.29 ± 1.76	95.92 ± 1.22	**100.00 ± 0.00**
6	79.39 ± 4.15	82.97 ± 3.98	92.31 ± 1.25	*94.83 ± 2.19*	89.46 ± 1.60	**96.84 ± 1.17**	83.12 ± 4.97	81.39 ± 2.87	94.26 ± 1.18
7	77.14 ± 2.69	72.08 ± 6.61	87.72 ± 1.39	86.42 ± 3.42	*88.17 ± 4.32*	83.48 ± 5.89	76.48 ± 5.33	85.44 ± 3.28	**93.21 ± 0.97**
8	55.88 ± 6.01	61.62 ± 9.81	57.02 ± 5.89	53.05 ± 8.71	60.26 ± 3.37	76.15 ± 4.37	*82.91 ± 1.39*	77.02 ± 4.91	**89.52 ± 2.82**
9	70.11 ± 2.93	72.89 ± 0.23	80.30 ± 3.76	84.04 ± 4.63	*86.32 ± 2.67*	82.17 ± 1.78	75.92 ± 3.44	79.39 ± 2.76	**90.16 ± 0.76**
10	46.91 ± 3.52	53.29 ± 2.92	51.79 ± 6.18	45.14 ± 8.77	50.29 ± 7.32	86.85 ± 8.32	87.22 ± 2.97	*90.31 ± 4.72*	**98.67 ± 0.88**
11	50.97 ± 2.61	52.08 ± 7.68	62.36 ± 3.27	61.85 ± 9.39	66.50 ± 6.30	88.57 ± 5.06	81.32 ± 1.52	*89.32 ± 2.60*	**94.02 ± 0.82**
12	70.82 ± 3.28	77.52 ± 3.56	80.15 ± 2.32	*84.40 ± 2.93*	80.02 ± 3.10	78.64 ± 4.79	76.26 ± 3.86	83.77 ± 2.93	**93.62 ± 1.25**
13	76.39 ± 2.88	82.32 ± 2.18	*87.29 ± 1.92*	84.14 ± 3.12	80.39 ± 2.66	75.62 ± 6.93	77.39 ± 4.24	83.81 ± 4.86	**89.92 ± 1.02**
14	85.67 ± 4.62	90.60 ± 1.82	94.03 ± 1.56	96.03 ± 0.73	92.30 ± 1.92	*99.45 ± 0.44*	91.62 ± 1.40	90.35 ± 0.63	**100.00 ± 0.00**
15	92.22 ± 3.71	90.08 ± 2.13	88.32 ± 1.78	93.45 ± 2.41	94.07 ± 0.91	**98.03 ± 1.07**	90.31 ± 2.35	95.11 ± 1.66	*96.26 ± 0.84*
OA (%)	75.67 ± 2.31	77.39 ± 2.57	80.32 ± 2.09	82.10 ± 1.21	80.67 ± 1.32	*89.31 ± 1.00*	84.77 ± 1.42	88.06 ± 0.82	**94.12 ± 0.87**
AA (%)	73.62 ± 2.64	76.43 ± 1.92	81.59 ± 2.31	79.21 ± 1.02	81.23 ± 1.76	*90.33 ± 1.06*	85.39 ± 1.68	87.61 ± 1.28	**95.07 ± 0.55**
Kappa	73.02 ± 1.86	77.16 ± 1.82	79.37 ± 1.85	77.61 ± 1.19	79.32 ± 1.53	*88.44 ± 1.08*	84.06 ± 1.83	87.26 ± 0.93	**94.66 ± 0.56**

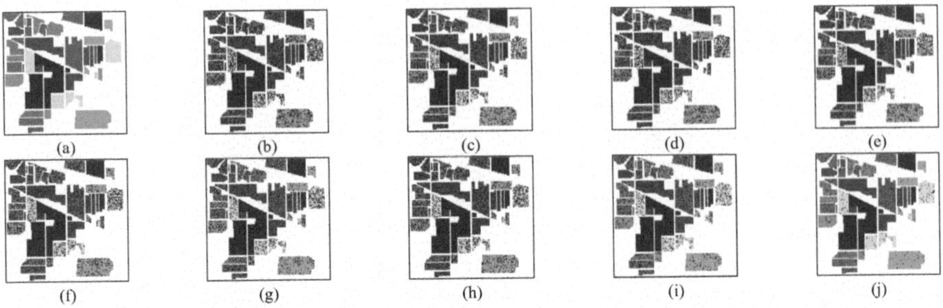

Fig. 7.7 Classification maps of comparison methods on IP dataset. **a** Ground-truth map; **b** AB-LSTM (OA = 56.68%); **c** SVM-RBF (OA = 66.96%); **d** 2D CNN (OA = 65.83%); **e** CRNN (OA = 66.28%); **f** miniGCN (OA = 68.73%); **g** S^{2}GCN (OA = 75.24%); **h** SF (OA = 65.24%); **i** SSFTT (OA = 82.96%); **j** S^{2}GFormer (OA = 93.55%)

contains many misclassification pixels in Corn-notill, Corn-mintill, and Soybean-mintill. The above experimental results evaluate that the effectiveness of our method is better than that of competitors in the IP scene.

Classification results on Salinas dataset: The Second experiment is conducted on Salians dataset to assess the classification performance of our S^{2}GFormer for medium-scale HSI datasets with "different objects with the same spectrum" (C8 and C15). The quantitative and corresponding qualitative classification results are summarized in Table 7.4 and Fig. 7.8, respectively.

From the quantitative classification results in Table 7.4, due to the Salinas dataset contains less interference noise, we can conclude that the classification results achieved by investigated methods on the Salinas dataset have been greatly improved compared with the results on the IP dataset. In addition, as a traditional classification

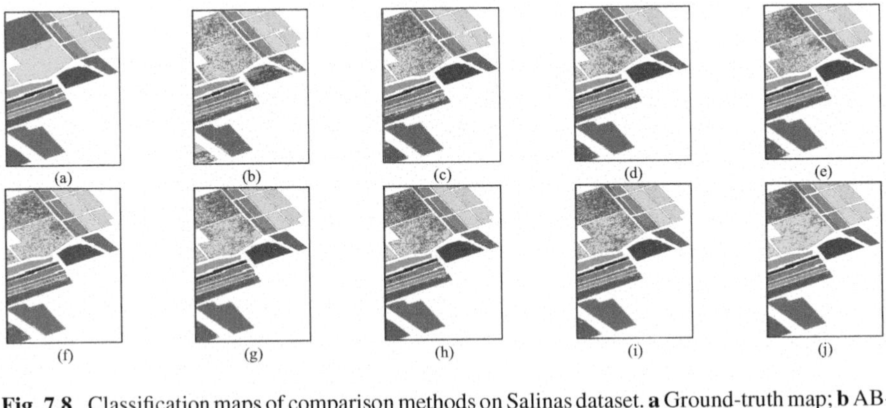

Fig. 7.8 Classification maps of comparison methods on Salinas dataset. **a** Ground-truth map; **b** AB-LSTM (OA = 74.83%); **c** SVM-RBF (OA = 86.31%); **d** 2D CNN (OA = 87.59%); **e** CRNN (OA = 87.64%); **f** miniGCN (OA = 86.75%); **g** S^{2}GCN (OA = 88.39%); **h** SF (OA = 88.80%); **i** SSFTT (OA = 90.67%); **j** S^{2}GFormer (OA = 96.92%)

method, the classification accuracies of SVM-RBF are 74.83, 83.89, and 72.21% in terms of OA, AA, and κ, there is no obvious difference between this result and the comparative in-depth learning method, showing that the traditional method has good classification performances for specific datasets. Due to the training sample deficiency, the CNN-based classification methods achieve unsatisfactory classification performance. For example, the OA, AA, and κ of CRNN are 87.64, 94.55, and 86.72%, respectively. Surprisingly, the classification accuracy of the GCN-based method for the Salinas dataset has not been significantly improved, indicating that the miniGCN and S^{2}GCN are unable to deal with "different objects with the same spectrum" very well. Specifically, the classification accuracy of S^{2}GCN in C8 (Grapes untrained) and C15 (Vineyard-untrained) are 70.68 and 70.36%. Thanks to the Transformer's self-attention mechanism, it can well extract the non-local spatial-spectral features of HSIs. The OA, AA, and κ, achieved by SSFTT are 90.67, 95.34, and 89.64%, respectively, which are the second-best performance. The results of our S^{2}GFormer are 96.92, 98.23, and 95.62%, which are improved by 6.25, 2.89, and 5.98% compared with these of SSFTT. This is mainly because our S^{2}GFormer method comprehensively utilizes the advantages of GCN-based methods and Transformer-based methods, thus, local spatial-spectral features are maintained while extracting non-local features. As the visual classification maps in Fig. 7.8, the visual classification map of our proposed method is closer to the Ground-truth map and reduces the misclassification rate significantly.

Classification results on UH2013 dataset: The third experiment is conducted on the UH2013 dataset to assess the classification performance of our S^{2}GFormer for large-scale HSI datasets. The quantitative and corresponding qualitative classification results are summarized in Table 7.5 and Fig. 7.9, respectively.

From the quantitative results demonstrated in Table 7.5, similar to the IP and Salinas datasets, the classification performance of our proposed S^{2}GFormer method is superior to that of competitive methods among all key indicators, showing the good classification ability of our method for large-scale datasets. Specifically, the results

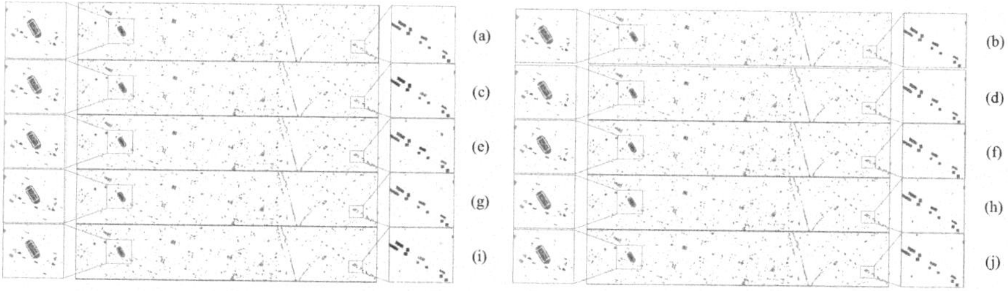

Fig. 7.9 Classification maps of comparison methods on UH2013 dataset. **a** Ground-truth map; **b** AB-LSTM (OA = 75.67%); **c** SVM-RBF (OA = 77.39%); **d** 2D CNN (OA = 80.32%); **e** CRNN (OA = 82.10%); **f** miniGCN (OA = 80.67%); **g** S^{2}GCN (OA = 89.31%); **h** SF (OA = 84.77%); **i** SSFTT (OA = 88.06%); **j** S^{2}GFormer (OA = 94.12%)

are improved by 4.81, 4.74, and 6.22% compared with the second-best method. Moreover, compared with GCN-based methods, i.e., miniGCN and S²GCN, Transformer-based methods, i.e., SF and SSFTT, do not obtain satisfactory advantages. Specifically, the OAs of the SF and SSFTT are only 84.77 and 88.06%, respectively. Furthermore, SC is suffering from out-of-memory during clustering. Furthermore, we should note that the S²GCN achieves the second-best results in all investigated classifiers, which indicates that the GCN mechanism contains a good advantage for HSI classification. Due to the high-level features for HSI classification, the deep classifiers (i.e., 2D CNN, CRNN, miniGCN, S²GCN, SF, and SSFTT) obtain remarkable performances. Specifically, the classification accuracies of deep classifiers in terms of the OA are increased by 4.65 and 2.93%. As for the visual performance comparison of all investigated methods on the UH2013 dataset is presented in Fig. 7.9. Our proposed S²GFormer achieves the best visual classification result among all methods, which confirms the advantage of our S²GFormer.

7.4.3 *Classification Visualization*

In this part, the classification results are visualized in Fig. 7.10 by embedding t-SNE [12] into S²GFormer and SSFTT to enhance the interpretability.

From the visualization, we can get the following conclusions: (1) the variance between the class center and the classification target is reduced compared with the original HSI distribution; (2) misclassifications are greatly reduced by S2GFormer; (3) the nodes of the same class are more clustered, and the centers of different classes are farther away in the classification maps of S²GFormer. In short, the feature representation extracted by S²GFormer is inter-class distance and intra-class compression compared with the competitive classifiers.

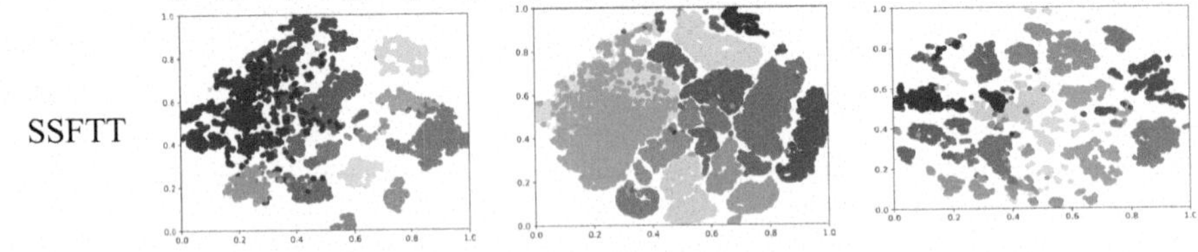

Fig. 7.10 Feature presentation for S²GFormer and SSFTT on three datasets

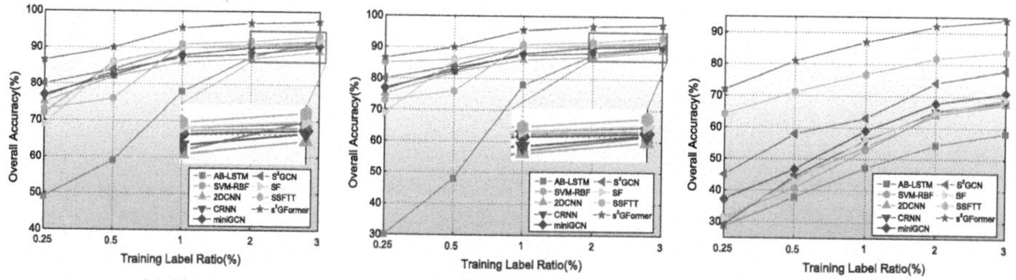

(a) IP (b) Salinas (c) UH2013

Fig. 7.11 The overall accuracies (%) of the investigated classifiers with the different training label ratio on three dataset. **a** IP, **b** Salinas, **c** UH2013

7.4.4 Influence of Label Ratio

To further evaluate the classification ability of the proposed method with a different number of training samples, we randomly select different labelled samples for investigated classifiers training from 0.25 to 3% on three datasets. The other experimental setups are shown in Sect. 7.4.1. The OA index is adopted to report the performances, and the results are shown in Fig. 7.11. Unsurprisedly, the performances of all classifiers are improved with the ratio of training samples increasing. Most significantly, the classification results of our proposed method are the best, no matter the number of training samples changes. Meanwhile, we can also observe that S^{2}GFormer has the best adaptability to small samples, and the volatility of the results is the smallest, verifying the progressiveness of the method design. Moreover, GCN-based classifiers can learn the information of neighbor nodes, and Transformer-based methods contain a self-attention mechanism to extract the global features of HSI, the two kinds of methods perform well in the condition of small samples, while the effect is not obvious as the number of training samples increasing. This result shows that only extracting local or global spatial-spectral features of HSI cannot effectively improve classification accuracy and adaptability of classifier.

7.4.5 Hyperparameter Analysis

In our S^{2}GFormer, we should preset five main hyperparameters (i.e., T, P_n, L, V_d, and V_h), and their optimization values are shown in Table 7.1. In this section, we explain the selection process of each hyperparameter and analyze the sensitivity of the proposed method to the hyperparameter when classifying. To tune the optimal weights, a grid search strategy is employed, and the indicator OA is utilized to record the experimental results.

The number of epochs T and learning rate L control the number of cycles of the algorithm and the step-size of parameter optimization, respectively. Therefore, we conducted some experiments to access the sensitivity of two paraments varying T

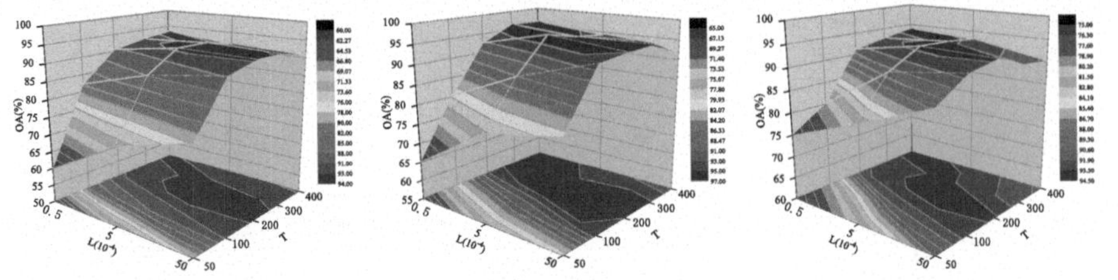

(a) IP (b) Salinas (c) UH2013

Fig. 7.12 Hyperparameter sensitivity analysis about the number of epochs T and learning rate L on IP (**a**), Salinas (**b**), and UH2013 (**c**)

from 50 to 400 and L from 5×10^{-5} to 5×10^{-3}. In addition, the other paraments are fixed as in Table 7.1. From the results in Fig. 7.12, we conclude that (1) the model can train faster with a bigger L, but it is not easy to find the optimal parameters; (2) a larger T can make the model more fully trained, but it is easy to lead to overfitting. A smaller T will lead to insufficient optimization of the algorithm parameters. Therefore, it is very important to select the appropriate T. (3) T and L are closely related. The selection of one parameter affects the decision of the other. To fully optimize the parameters, while ensuring the training efficiency, it is appropriate to select the value of T and L as 200 and 5×10^{-4} from the result analysis.

The patch size P_n is closely related to the computational complexity and the amount of spatial information preserved by the patch. In this chapter, we investigated the different P_n impacts on HSI classification, and the results are shown in Fig. 7.13a. In general, larger patches can preserve more local spatial information, while, accompanied by more noise, the classification accuracy is reduced. On the contrary, a smaller patch size will increase the number of patches, increasing the computational complexity of the algorithm. From Fig. 7.13a, we note that the proposed method achieves the highest classification accuracy, while the patch size is the smallest with a 9×9 patch size. The self-attention head can extract global information from the image. After that, we analyzed the effect of the number of self-attention heads in S^{2}GFormer on HSI classification. From the results in Fig. 7.13b, we can observe that the classification accuracy of the algorithm is the highest as $V_h = 4$. The results indicate that excessive to extract the global feature information of HSI is not beneficial to improve the classification accuracy of the classifier.

7.4.6 *Ablation Studies*

The proposed S^{2}GFormer mainly contains three modules: including pixel to region transform module, patch-wise spectral embedding block, and multi-layer Graphormer Encoder module. And each novel component plays a pivotal role in

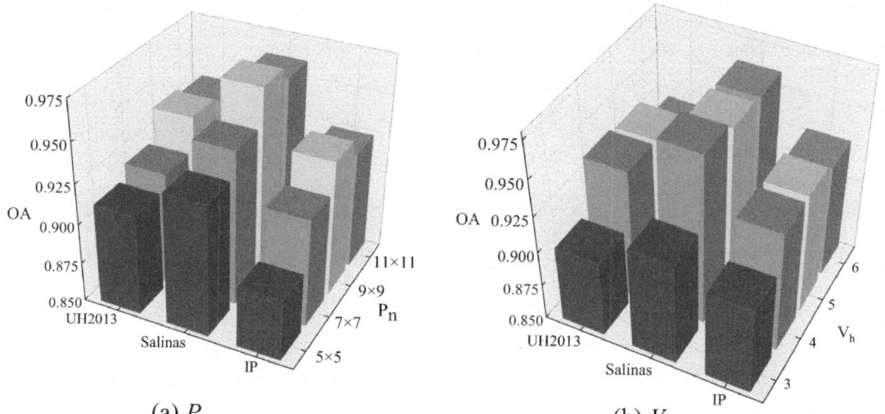

(a) P_n (b) V_h

Fig. 7.13 Hyperparameter sensitivity analysises on three datasets. **a** patch size P_n. **b** Number of heads V_h

the proposed method. In this part, to see how they contribute to the overall classification performance, we conduct a set of ablation studies. Thus, S^{2}GFormer without the pixel-to-region transform module is called S^{2}GFormer-V$_1$. Besides, S^{2}GFormer-V$_2$ is obtained by replacing the patch-wise spectral embedding block with the mean operation. Moreover, the S^{2}GFormer-V$_3$ is obtained by replacing the Graphormer Encoder module with a conventional Transformer. Furthermore, S^{2}GFormer-V$_4$ is formed by removing the Residual block in the Graphormer Encoder. The OA, AA, and κ are employed to report the classification results, as shown in Fig. 7.14. From the result, we can observe that (1) compared with the S^{2}GFormer, the classification accuracies of all compared methods are decreased in all indicators; (2) the index declines in the simplified methods are different, which means that each module designed in the proposed method has a different influence on the classification results; (3) all modules proposed in S^{2}GFormer contribute to classification accuracy improving.

In addition, the role of graph residual block in Graphormer Encoder is investigated. Specifically, the residual block is removed in different Graphormer Encoders on the IP dataset. The results are shown in Table 7.6, which indicates that graph residual block

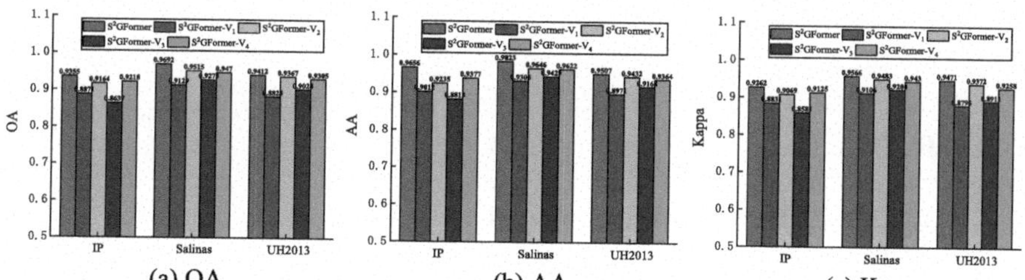

(a) OA (b) AA (c) Kappa

Fig. 7.14 Ablation experimental results on IP, Salinas, UH2013. **a** OA, **b** AA, **c** Kappa

Table 7.6 Ablation study of encoder in graphormer with graph RES

Encoder1	Encoder2	Encoder3	OA (%)
✗	✗	✗	92.18
✔	✗	✗	92.67
✗	✔	✗	93.19
✗	✗	✔	92.86
✔	✔	✔	**93.55**

Block for HSI classification on IP dataset. ✔—Add RES. block in to the specified encoder. ✗—No RES. block in the specified encoder

improves the classification accuracy of the algorithm to a certain extent compared with a base graph convolution layer.

7.5 Conclusions

This chapter proposes a unique Graphormer with spatial-spectral features for HSI classification. The design of Follow Patch mechanism can transform the pixel in HSI to locality-preserving regions, and the computational complexity of S^2GFormer can be reduced. After that, a patch-wise spectral embedding block is developed to extract comprehensive spectral information for patches with a newly designed neighborhood convolution. Finally, the multi-layer Graphormer encoder module, combing graph convolution, and Transformer structures into a unified Transformer, is proposed to extract the representative spatial-spectral features from the patch for HSI classification. In this study, the three aforementioned parts are jointly integrated into a unified end-to-end network, and each component benefits the other. Experimental results on three commonly used HSI data sets demonstrate that our S^2GFormer method achieves better performance than other state-of-the-art classifiers.

In the future, we will apply more techniques, e.g., self-supervised learning and reinforcement learning, to Graphormer to improve its performance on the HSI classification task. Moreover, we will establish more integration forms of graph convolution and Transformer, and explore the implementation of graph node coding in the Transformer coding.

References

1. Y. Ding, Z. Zhang, X. Zhao, W. Cai, N. Yang, H. Hu, X. Huang, Y. Cao, W. Cai, Unsupervised self-correlated learning smoothy enhanced locality preserving graph convolution embedding clustering for hyperspectral images. IEEE Trans. Geosci. Remote Sens. **60**, 1–16 (2022)

2. D. Hong, L. Gao, J. Yao, B. Zhang, A. Plaza, J. Chanussot, Graph convolutional networks for hyperspectral image classification. IEEE Trans. Geosci. Remote Sens. **59**(7), 5966–5978 (2020)
3. Y. Ding, X. Zhao, Z. Zhang, W. Cai, N. Yang, Graph sample and aggregate-attention network for hyperspectral image classification. IEEE Geosci. Remote Sens. Lett. **19**, 1–5 (2022)
4. Z. Chen, Z. Lu, H. Gao, Y. Zhang, J. Zhao, D. Hong, B. Zhang, Global to local: a hierarchical detection algorithm for hyperspectral image target detection. IEEE Trans. Geosci. Remote Sens. **60**, 1–15 (2022)
5. J.D.M.-W.C. Kenton, L.K. Toutanova, *BERT: Pre-training of Deep Bidirectional Transformers for Language Understanding*, pp. 4171–4186
6. Z. Zhang, Y. Ding, X. Zhao, L. Siye, N. Yang, Y. Cai, Y. Zhan, Multireceptive field: an adaptive path aggregation graph neural framework for hyperspectral image classification. Exp. Syst. Appl. **45**, 119508 (2023)
7. Y. Chen, H. Jiang, C. Li, X. Jia, P. Ghamisi, Deep feature extraction and classification of hyperspectral images based on convolutional neural networks. IEEE Trans. Geosci. Remote Sens. **54**(10), 6232–6251 (2016)
8. L. Mou, P. Ghamisi, X.X. Zhu, Deep recurrent neural networks for hyperspectral image classification. IEEE Trans. Geosci. Remote Sens. **55**(7), 3639–3655 (2017)
9. A. Qin, Z. Shang, J. Tian, Y. Wang, T. Zhang, Y.Y. Tang, Spectral–spatial graph convolutional networks for semisupervised hyperspectral image classification. IEEE Geosci. Remote Sens. Lett. **16**(2), 241–245 (2018)
10. D. Hong, Z. Han, J. Yao, L. Gao, B. Zhang, A. Plaza, J. Chanussot, SpectralFormer: rethinking hyperspectral image classification with transformers. IEEE Trans. Geosci. Remote Sens. (2021)
11. L. Sun, G. Zhao, Y. Zheng, Z. Wu, Spectral–spatial feature tokenization transformer for hyperspectral image classification. IEEE Trans. Geosci. Remote Sens. **60**, 1–14 (2022)
12. L. Van der Maaten, G. Hinton. Visualizing data using t-SNE. J. Mach Learn. Res. **9**(11) (2008)

Chapter 8
Self-Supervised Locality Preserving Low-Pass Graph Convolutional Embedding for Large-Scale Hyperspectral Image Clustering

8.1 Introduction

Hyperspectral imagery (HSI) acquired by remote sensing system are composed of hundreds of contiguous and narrow spectral bands with abundant spatial-spectral information in the electromagnetic spectrum [1–3]. Due to its unique advantages, HSI has attracted lots of attention and has been widely applied in various fields, including military reconnaissance, urban mapping, biochemical detection, forest fire detection, and target recognition [4–6], etc.

As one of the fundamental and common techniques for HSI practical application, hyperspectral imagery classification (HSIC), aiming to divide each pixel contained in the image into different landcover, has been widely studied. Recently, many supervised or semi-supervised deep learning HSIC methods have achieved promising performances [7, 8]. However, these methods rely on annotated samples to train the network. In practice, it is a high labor-intensive and time-consuming progress to obtain the labeled HSI pixels, which is a problem that must be considered in the application of the HSIC method. To solve the aforementioned problem, the unsupervised clustering method, aiming to assign similar pixels into unique land-cover classes with no prior labels, has attracted increasing attention. Compared with supervised or semi-supervised methods, HSI clustering is often a fundamental but challenging task, due to prior knowledge deficiency, large spectral variability, and high-dimension of HSI [9, 10].

In the past several decades, various clustering methods have been designed for HSI clustering [11–14]. In the early stage, some traditional clustering algorithms developed for natural images have been applied to HSI clustering, such as k-means [15], spectral clustering (SC) [16], fuzzy c-means (FCM) [17], independent component analysis model (ICA) [18], etc. However, these algorithms only extract spectral information, the spatial information contained in HSI is ignored, which limits their performances. Thus, to fully utilize the spectral-spatial information of HSI and improve the clustering accuracy, some spectral-spatial HSI clustering methods

Y. Ding et al., *Graph Neural Network for Feature Extraction and Classification of Hyperspectral Remote Sensing Images*, Intelligent Perception and Information Processing, https://doi.org/10.1007/978-981-97-8009-9_8

[19–21] are designed, and better performances have been achieved than that of only spectral-based methods. For example, Elhamifar and Vidal [22] proposed FCM-S1, in which the spatial-contextual information is incorporated. Furthermore, to solve the high-dimensional problem of HSI, some sparse subspace clustering (SSC) algorithms [23–25] have appeared, which can cluster the pixels in the low-dimensional subspace by learning the coefficient representation matrix and constructing the affinity matrix. In addition, Zhai et al. [26] integrated the total variation and collaborative subspace clustering (CSC) into a unified framework to present the spectral-spatial information of HSIs. In [27], a scalable graph-based clustering method named SGCNR was designed by Wang et al. using nonnegative relaxation to present HSI features.

Unfortunately, the aforementioned methods are shallow classifiers based on the hand-crafted feature or raw feature spaces, which cannot learn the high-level features of HSI and are not suitable for characterizing the large spectral variability and the complex internal structure of HSI datasets effectively [28–30]. Thus, some deep learning techniques have been applied to overcome the limitation. With the successful application of deep learning-based clustering in natural image clustering, some deep learning clustering methods have been proposed to tackle the HSI clustering task. We can divide most of the available deep HSI clustering methods into two steps, including high-level feature extraction adopting deep learning techniques and traditional clustering. In [31], a graph regularized residual subspace clustering network (GR-RSCNet) was introduced, where the robust nonlinear affinity and deep spectral-spatial representation were jointly learned to present the complex features of HSI in a deep neural network. In Cai et al. [32], presents a neighborhood contrastive subspace clustering (NCSC) network, a scalable and robust DSC approach, for unsupervised classification of large hyperspectral images. Inspired by the auto-encoder technique in the computer vision field, some researchers have introduced auto-encoder into the clustering method for spatial-spectral feature learning. For instance, a spatial-spectral subspace clustering network (DS^{3}C-Net) was designed by Lei et al. [33] to extract spatial-spectral features by collaborative constraint and the multi-scale auto-encoder. Hou et al. [34] introduced contrast learning [35] into HSI features learning, and the unlabeled samples were utilized to alleviate the label information insufficient problem. Nevertheless, the aforementioned methods are not designed specifically for large-scale HSI clustering, and the clustering performances on complex large-scale HSI dataset clustering should be improved. In addition, due to labeled data deficiency, large spectral variability, and high-dimension of HSIs, most of exist deep clustering methods cannot effectively learn remarkable and adaptive high-level features for HSI clustering, which leads to poor clustering accuracy. Finally, most of the deep clustering methods ignore the layer-wise local features, which results in poor adaptability to different HSI datasets.

To address these drawbacks, we propose a novel self-supervised locality preserving low-pass graph convolutional embedding method (L2GCC) for large-scale hyperspectral image clustering. Specifically, a spectral-spatial transformation HSI pre-processing mechanism is introduced to learn superpixel-level spectral-spatial features from HSI and reduce the number of graph nodes for subsequent network processing. In addition, we design a low-pass graph convolutional embedded autoencoder to learn a hidden representation of the graph by training the inner product decoder to reconstruct the graph, with which a smoother graph compact representation can be learned. Furthermore, we proposed a layer-wise graph attention module to preserve the local features of each layer and improve the clustering adaptability to different HSI datasets. Finally, we develop a self-training strategy, in which a self-training clustering objective employs soft labels to supervise the clustering process and obtain appropriate hidden representations for node clustering. In L2GCC, we jointly integrate locality preserving low-pass graph convolution autoencoder and self-training clustering into a single end-to-end network to achieve mutual benefit by adopting jointly optimizing loss during network training. In summary, the main contributions of L2GCC are summarized as follows:

- We design a locality preserving low-pass graph convolutional embedding autoencoder, in which the low-pass graph convolution and layer-wise graph attention are designed to extract the smoother features and locality features, respectively.
- We indoduce a self-training clustering mechanism to refine the clustering results and solve the labeled samples deficiency problem of HSIs, and soft labels generated from graph embedding are proposed to supervise the self-training clustering process.
- The locality preserving low-pass graph convolutional autoencoder presents self-training clustering are jointly integrated into a single network to achieve mutual benefit by adopting jointly optimizing loss. Compared with state-of-the-art clustering baselines, the experimental results verify our L2GCC superiority.

8.2 Background

8.2.1 Notations and Descriptions

In this paper, all vectors are represented in lowercase and bold, all matrices are denoted in uppercase. The n, d, and c are adopted to represent the size of nodes, dimension of HSI (node-set number of clusters). The input node-set $\{x_i\}_{i=1}^{n}$ can be expressed as

$$\boldsymbol{X} = [x_1, x_2, \cdots, x_n] \in \mathbb{R}^{n\times d} \tag{8.1}$$

Let $\mathcal{V}$ be the nodes set, $\mathcal{E}$ be the edge set between nodes, $\boldsymbol{X} \in \mathbb{R}_{n\times d}$ be the node feature matrix, then the graph can be denoted as $\mathcal{G} = (\mathcal{V}, \mathcal{E})$, in which the adjacency matrix is denoted as $\boldsymbol{A} \in R_{n\times n}$, and the corresponding degree matrix is presented as $\boldsymbol{D} \in \mathbb{R}_{n\times n}$. Furthermore, $\cdot\|\cdot$ denotes the concatenation operation. The main notations adopted in the paper and their descriptions are demonstrated in Table 8.1.

Table 8.1 Main notions and descriptions

Notations	Description
n	Number of nodes contained in the graph
d_i	The dimension of $\boldsymbol{X}$ and the ith latent feature
c	Number of clusters
W	Learnable weight matrix
$\boldsymbol{S}_i$	ith superpixel
$\boldsymbol{X}$	Node feature matrix
$\mathcal{G} = (\mathcal{V}, \mathcal{E})$	Graph $\mathcal{G}$ with edges set $\mathcal{E}$ and nodes set $\mathcal{V}$
$\boldsymbol{A}, \widehat{\boldsymbol{A}}_{ij}$	Adjacency matrix of $\mathcal{G}$. The adjacency matrix of a reconstructed graph
$\boldsymbol{D} = diag(d_1, \cdots, d_n)$	Degree matrix of A
$\boldsymbol{L}, \boldsymbol{L}_{sym}$	Graph Laplacian Symmetric and normalized Laplacian of $\mathcal{G}$
$\boldsymbol{U}$	Eigenvector matrix of $\boldsymbol{L}_{sym}$
$\cdot\|\cdot$	Concatenation operation
$\boldsymbol{Z}_i$	ith layer encoder output
λ_{max}, λ_i	Largest eigenvalue, ith eigenvalue
$\mathcal{N}(\cdot)$	Neighbourhood of $\cdot$
$\odot$	Element-wise product
P, Q	The target distribution. The soft labels distribution
η	Learning rate

8.2.2 Convolution on Graph

We can explain the graph convolution operation as both spatial operators [36] and spectral operators [37]. In the paper, we define the convolution operator from the frequency, which presents the graph signals in the spectral domain. Given a graph $\mathcal{G} = (\mathcal{V}, \mathcal{E}, \boldsymbol{A})$, where $\mathcal{V} = \{v_i\}_{i=1,\cdots,n}$ denotes the set of the vertex with n nodes and m channels, $\mathcal{E} = \{e_{ij}\}_{i,j=1,\cdots,n}$ is the set of edges between node (superpixel), the adjacency matrix of the graph is $\boldsymbol{A} \in \mathbb{R}^{n\times n}$, and the weight of the edge between vertex i and vertex j is A_{ij}, in which $A_{ij} = 1, \textit{if}\, \{v_i, v_j\} \in \mathcal{E}$, indicating that there is an edge between node v_i and node v_i; otherwise $A_{ij} = 0$. The node feature matrix is presented as $\boldsymbol{X} = \{\boldsymbol{x}_1; \cdots; \boldsymbol{x}_n\}$, where $\boldsymbol{x}_i \in \mathbb{R}^m$ is the feature value associated with the node v_i, n denote the number of nodes contained in the graph, m is the dimension of ith latent feature. In the graph, we can denote the topological structure of the graph $\mathcal{G}$ by constructing an adjacency matrix $\boldsymbol{A}$. The mathematical representation of the normalized graph can be defined as $\boldsymbol{L} = \boldsymbol{I_n} - \boldsymbol{D}^{-\frac{1}{2}}\boldsymbol{A}\boldsymbol{D}^{-\frac{1}{2}}$, where $\boldsymbol{D}$ is the degree matrix, $\boldsymbol{D}_{ii} = \sum_j(\boldsymbol{A}_{i,j})$. The $\boldsymbol{L}$ is real symmetric positive semidefinite, and we can factor $\boldsymbol{L}$ as $\boldsymbol{L} = \boldsymbol{U}\boldsymbol{\Lambda}\boldsymbol{U}^T$, where $\boldsymbol{U} = \left[\boldsymbol{u}_0, \boldsymbol{u}_1, \cdots, \boldsymbol{u}_{n-1}\right] \in \mathbb{R}^{N\times N}$ is the eigenvectors matrix of $\boldsymbol{L}$, i.e., $\boldsymbol{U}\boldsymbol{U}^T = \boldsymbol{E}$, where $\boldsymbol{E}$ is the identity matrix, $\boldsymbol{\Lambda}$ is the diagonal matrix of eigenvalues, $\boldsymbol{\Lambda}_{ii} = \boldsymbol{\lambda}_i$, $\boldsymbol{\lambda}_i$ denotes the eigenvalue. Given a graph signal $\boldsymbol{X} \in \mathbb{R}^{N \times M}$ (graph node feature matrix), the Fourier transform of $\boldsymbol{X}$ can be expressed as $\mathcal{F}(\boldsymbol{X}) = \boldsymbol{U}^T\boldsymbol{X}$, and the inverse graph Fourier transform can be expressed as $\mathcal{F}^{-1}(\widehat{\boldsymbol{X}}) = \boldsymbol{U}\widehat{\boldsymbol{X}}$, where $\widehat{\boldsymbol{X}}$ is the result of Fourier transform. The graph Fourier transform projects the graph feature vector into the orthogonal space, in which the basis is composed of the eigenvector of the normalized graph Laplace operator. The elements in $\widehat{\boldsymbol{X}}$ are the coordinates projected by $\boldsymbol{X}$ in the orthogonal space, and the $\boldsymbol{X}$ can be expressed as $\boldsymbol{X} = \sum_i \hat{\boldsymbol{x}}_i \boldsymbol{u}_i$. We can express the graph convolution of $\boldsymbol{X}$ with a graph filter $\boldsymbol{g}$ as

$$\begin{aligned} \boldsymbol{g} * \boldsymbol{X} &= \mathcal{F}^{-1}(F(\mathrm{x}) \odot F(\mathrm{g})) \\ &= \boldsymbol{U}\big(\boldsymbol{U}^T\boldsymbol{X} \odot \boldsymbol{U}^T\boldsymbol{g}\big) \end{aligned} \tag{8.2}$$

If the graph filter is expressed as $\boldsymbol{g}_\theta = diag(\boldsymbol{U}^T\mathbf{g})$, then the Eq. (8.2) can be simplified as

$$\boldsymbol{g}_\theta * \boldsymbol{X} = \boldsymbol{U}\boldsymbol{g}_\theta(\boldsymbol{\Lambda})\boldsymbol{U}^{\mathrm{T}}\boldsymbol{X} \tag{8.3}$$

The spectral-based graph convolution algorithms all follow the definition of Eq. (8.3). The graph filter is important to spectral-based graph convolution algorithms, graph filter designing can determine the spectral-based graph convolution algorithms' performances. The aggregation operation in graph convolutional network (GCN) [38] is expressed as:

$$\boldsymbol{X}_{i+1} = \sigma\left(\boldsymbol{D}^{-\frac{1}{2}}\widehat{\boldsymbol{A}}\boldsymbol{D}^{-\frac{1}{2}}\boldsymbol{X}_i\boldsymbol{W}_i\right) \tag{8.4}$$

where $\boldsymbol{X}_i$ is output feature matrix of i-th layer, and $\boldsymbol{X}_0 = \boldsymbol{X}$. Equation (8.4) show the node feature transformation of GCN from $\boldsymbol{X}_i \in \mathbb{R}^{n \times c_i}$ to $\boldsymbol{X}_{i+1} \in \mathbb{R}^{n \times c_{i+1}}$. σ is a nonlinear activation function, $\boldsymbol{W}_i \in \mathbb{R}^{c_i \times c_{i+1}}$ is a learnable weight matrix to realize the linear transformation on the feature matrix. $\widehat{\boldsymbol{A}} = \boldsymbol{A} + \boldsymbol{I}, \boldsymbol{D} = diag(d_1, \cdots d_n,)$ is the degree matrix to performing normalization on $\widehat{\boldsymbol{A}}$.

8.2.3 Attention Convolution on Graph

Given a graph $\mathcal{G}$, the feature eigenvector of each node is $\boldsymbol{x}_i \in \mathbb{R}^m$. A new eigenvector $\boldsymbol{x}_i' \in \mathbb{R}^{m'}$ can be obtained by linear transformation for $\boldsymbol{x}_i$, we can express the operation process as

$$\boldsymbol{x}_i' = \boldsymbol{W}\boldsymbol{x}_i, \boldsymbol{W} \in \mathbb{R}^{m \times m'} \tag{8.5}$$

If the node j is a neighbor of node i, we can adopt the attention mechanism to calculate the importance of node j to node i, the attention score e_{ij} can be calculated as

$$e_{ij} = \left(\sigma\left(\boldsymbol{a}^T\left[\boldsymbol{W}\boldsymbol{x}_i || \boldsymbol{W}\boldsymbol{x}_j\right]\right)\right) \tag{8.6}$$

where || denotes concatenation operation, $\boldsymbol{a}^T \in \mathbb{R}^{2m'}$ is the parameter vector, σ is a nonlinear layer, i.e., *LeakyReLU*.

The e_{ij} can be converted to the following form through a softmax function

$$\alpha_{ij} = softmax\left(e_{ij}\right) = \frac{\exp\left(\sigma\left(\boldsymbol{a}^T\left[\boldsymbol{W}\boldsymbol{x}_i || \boldsymbol{W}\boldsymbol{x}_j\right]\right)\right)}{\sum_{k \in N_i} \exp\left(\sigma\left(\boldsymbol{a}^T\left[\boldsymbol{W}\boldsymbol{x}_i || \boldsymbol{W}\boldsymbol{x}_j\right]\right)\right)} \tag{8.7}$$

The mathematical expression of graph convolution output of graph attention network (GAT) [39] is

$$\boldsymbol{x}_i' = \sigma\left(\sum_{j \in n_i} \alpha_{ij} \cdot \boldsymbol{W}^T \boldsymbol{x}_j\right) \tag{8.8}$$

where σ is the activate function, a_{ij} is learned attention weight, n_i denote the number of neighbor nodes of the node i.

8.2.4 *Graph Auto-Encoder*

The graph autoencoder (GAE) [40] utilize multi-layer GCN to learn the hidden presentation of graph nodes embeddings, and the inner-products between two nodes are computed, and then they are mapped into a probability space to calculate the similarities through a sigmoid function, the graph is reconstructed by computing the graph reconstructed loss. The network architectures of GAEs [40–43] are often asymmetric, while we can employ graph sharpening to designs the symmetric architecture. For instance, adversarial learning is incorporated into GAE to enhance the robustness in Adversarial Regularized Graph Auto-Encoder [43]. In GAE with Adaptive Graph Convolution [42], the high-order convolution operator is introduced into GAE to improve the presentation capacity.

8.3 The Proposed Method

8.3.1 Problem Definition and Overall Framework

Given a non-directed graph $\mathcal{G}$, the goal is to classify the nodes into c clusters, i.e., $\{C_1, C_2, \cdots, C_c\}$, by using a self-supervised locality preserving low-pass graph convolutional embedding auto-encoder to learn the latent representation $\boldsymbol{Z}$. Therefore, the nodes in a single cluster contain similar spectral feature values and spatial structure.

From the flow-diagram of L2GCC in Fig. 8.1, we can partition our method into three parts, including pixel to region transformation, locality preserving low-pass graph convolutional embedding encoder, and graph convolution embedding decoder.

- **Pixel to Region Transformation:** The HSI is segmented into local superpixel, and the mean spectral value of pixels contained in the superpixel is treated as the feature value, with which the number of graph node is reduced.
- **Locality Preserving Low-pass Graph Convolutional Embedding Encoder:** The node feature and the graph structure are taken as the input of the network. The hidden representation $\boldsymbol{Z}$ is learned by the locality preserving low-pass graph convolutional embedding autoencoder, in which the low-pass GCN and layer-wise GAT are designed to extract the smoother features and locality features, respectively.
- **Graph Convolutional Embedding Decoder:** The proposed method contains two decoders. Namely, the graph is reconstructed by minimizing the KL divergence between the adjacency matrix $\boldsymbol{A}$ distribution and the adjacency matrix $\widehat{\boldsymbol{A}}$ of reconstructed graph distribution, and self-training clustering can be performed

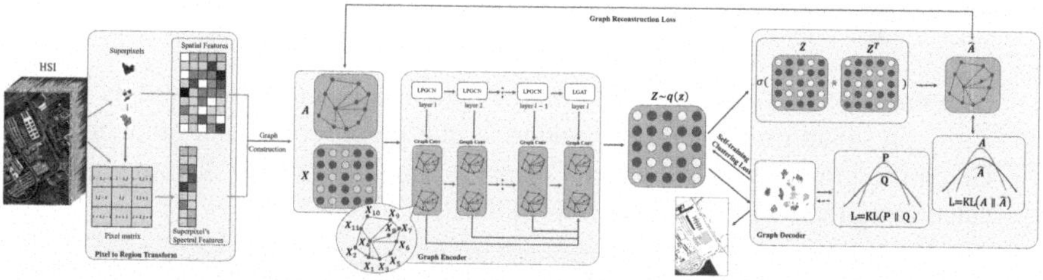

Fig. 8.1 The flow-diagram of the proposed hyperspectral image clustering network with self-supervised locality preserving low-pass graph convolutional embedding (L2GCC). Firstly, the HSI is transformed into graph nodes with pixel to region transformation module. Then the adjacency matrix $\boldsymbol{A}$ is calculated and the graph is constructed. Subsequently, we propose a locality preserving low-pass graph convolutional embedding autoencoder to learn the hidden representation $\boldsymbol{Z}$, in which the low-pass GCN and layer-wise GAT are designed to extract the smoother features and locality features, respectively. Finally, we introduce the self-training clustering to cluster the graph nodes, and the weights are updated by jointly optimizing loss during network training

by minimizing the KL divergence between the target distribution $\boldsymbol{P}$ and the soft label's distribution $\boldsymbol{Q}$.

The HSI preprocessing, graph convolutional embedding autoencoder, and node clustering are jointed in a unified network, and each component benefits the others. Therefore, the proposed network is an end-to-end training network.

8.3.2 Locality Preserving Low-Pass Graph Convolutional Embedding Encoder

To partition the graph nodes into different clusters, the basic assumption is that the nodes in the same cluster are supposed to be closely related, that is, they have similar spatial-spectral features. Therefore, the latent representation $\mathbf{Z}$ should be smooth on the graph manifold.

Assumption 1. In the ideal case, the distance within clusters is the smallest and the distance between clusters is the largest, namely

$$\begin{cases} \min \sum_{i=1}^{m} \sum_{j=1}^{m} d(i,j) = \min \sum_{i=1}^{m} \sum_{j=1}^{m} f(x_i) - f(x_j)_2^2 \\ \max \sum_{\alpha=1}^{c} \sum_{\beta=1}^{c} d(\alpha,\beta) = \max \sum_{\alpha=1}^{c} \sum_{\beta=1}^{c} y(c_\alpha) - y(c_\beta)_2^2 \end{cases} \tag{8.9}$$

where $\boldsymbol{x}_i$ is the node feature of node i, $f(\cdot)$ is the mapping that attempts to extract node's optimal spatial-spectral representation, m in the number of nodes contained in a cluster, $y(\cdot)$ is the function to calculate the central feature of the cluster, c is the number of clusters (classes).

(1) *Low-pass Graph Convolutional Filter:* A graph signal (node feature) is smooth if the presentation of neighbor nodes on the graph contains similar features. Given the graph signal $\boldsymbol{X}$, we can measure the graph signal $\boldsymbol{X}$ smoothness by calculating the Rayleigh quotient [44], i.e.,

$$R(\boldsymbol{L}, \boldsymbol{X}) = \frac{\boldsymbol{X}^{\mathrm{T}}\boldsymbol{L}\boldsymbol{X}}{\boldsymbol{X}^{\mathrm{T}}\boldsymbol{X}} = \frac{\sum_{(x_i,x_j)\in\mathcal{E}} (x_i - x_j)^2}{\sum_{i\in\mathcal{V}} x_i^2} \tag{8.10}$$

where $\boldsymbol{L}$ is the graph Laplacian, x_i is the i-th node feature. As stated by Eq. (8.10), smooth signals are supposed to have similar values on nearby nodes, and the smoother signals contain a lower Rayleigh quotient.

Theorem 1. *The smoother signals can be obtained by filtering high-frequency basis signals and preserving low-frequency basis signals.*

Following the principle of Theorem 1, we can design a low-pass graph filter to process the signals of a graph for node clustering. We can express a low-pass graph filter in many forms. In this paper, the frequency response function of low-pass graph filter is

$$p(\lambda_i) = 1 - k\lambda_i \tag{8.11}$$

where k is a real value, λ_i denotes the ith eigenvalues. Referring to the frequency response function $p(\cdot)$ in Eq. (8.11), the mathematical of graph filter $\boldsymbol{G}$ is expressed as

$$\boldsymbol{G} = \boldsymbol{U}p(\boldsymbol{\Lambda})\boldsymbol{U}^{-1} = \boldsymbol{U}(I - k\boldsymbol{\Lambda})\boldsymbol{U}^{-1} = I - k\boldsymbol{L} \tag{8.12}$$

Theorem 2. *Given a graph filter* $\boldsymbol{G}$*, if its frequency response function* $p(\lambda)$ *is nonnegative and nonincreasing for all* λ_i*, then given any signal* $\boldsymbol{X}$ *and the filtered* $\tilde{\boldsymbol{X}} = \boldsymbol{G}\boldsymbol{X}$*, we always have.*

$$R(\tilde{\boldsymbol{X}}, \boldsymbol{X}) \leq R(\boldsymbol{L}, \boldsymbol{X}) \tag{8.13}$$

According to Theorem 2, the filtered node features contain similar feature values in each dimension if such a low-pass graph filter is applied on graph signal, which is easier for subsequent network clustering. Perform graph convolution on the node feature matrix $\boldsymbol{X}$, the filtered feature $\tilde{\boldsymbol{X}}$ can be obtained as

$$\tilde{\boldsymbol{X}} = \boldsymbol{G}\boldsymbol{X} = \boldsymbol{U}p(\boldsymbol{\Lambda})\boldsymbol{U}^{-1} \cdot \boldsymbol{U}\boldsymbol{p} = \sum_{i=1}^{n}(1 - k\lambda_i)c_i\boldsymbol{u}_i = \sum_{i=1}^{n} c_i'\boldsymbol{u}_i \tag{8.14}$$

where $\tilde{\boldsymbol{X}} = [\tilde{\boldsymbol{x}}_1, \tilde{\boldsymbol{x}}_2, \cdots, \tilde{\boldsymbol{x}}_n]^T$, c_i is the coefficient of i-th eigenvector $\boldsymbol{u}_i$.

And the l-order graph convolution is defined as

$$\tilde{\boldsymbol{X}} = (I - k\boldsymbol{L})^l\boldsymbol{X} = \tilde{\boldsymbol{G}}\,\boldsymbol{X} \tag{8.15}$$

In Eq. (8.15), $\tilde{\boldsymbol{G}} = (I - k\boldsymbol{L})^l$ is the filter of l-order graph convolution, and its frequency response function is

$$p(\lambda_q) = \left(1 - k\lambda_q\right)^l \tag{8.16}$$

In practice, the graph Laplacian $\boldsymbol{L}$ is often normalized as

$$\tilde{\boldsymbol{L}}_{sym} = \tilde{\boldsymbol{D}}^{-\frac{1}{2}}\tilde{\boldsymbol{L}}\tilde{\boldsymbol{D}}^{-\frac{1}{2}} = \tilde{\boldsymbol{U}}\tilde{\boldsymbol{\Lambda}}\tilde{\boldsymbol{U}}^{-1} \tag{8.17}$$

Here, $\tilde{L}$ and $\tilde{D}$ are Laplacian matrix and degree matrix corresponding to $\tilde{A} = I + A$, the $\tilde{\Lambda}$ is decomposition of $\tilde{L}_{sym}$. Then we can express the filter as

$$G = I - k\tilde{L}_{sym} \tag{8.18}$$

It should be emphasized that if the k is set as 1, the graph filter is the traditional GCN filter. Equation (8.18) introduces the idea of designing a low-pass convolution kernel, however, different k values will lead to a different performance of the low-pass filter. Referring to the proof of Theorem 1 and Eq. (8.17), the filtered graph signal $\tilde{X}$ smoothness can be measured by

$$R\left(L, \tilde{X}\right) = \frac{\tilde{X}^T L \tilde{X}}{\tilde{X}^T \tilde{X}} = \frac{\sum_{i=1}^{n} g_i \lambda_i}{\sum_{i=1}^{n} g_i} \tag{8.19}$$

where $g_i > 0$ is a coefficient. To obtain a smoother filter, the g_i should decrease as λ_i increases. The maximum eigenvalue is denoted by λ_{max}. Theoretically, the graph filter is not low-pass, if $k > 1/\lambda_{max}$, because g_i increase in $\left(1/k, \lambda_{max}\right)$. However, not all high-frequency basis signals can be filtered as $k < 1/\lambda_{max}$. Therefore, the optimal value of k is $1/\lambda_{max}$.

Referring to the propagation rules of GCN, the propagation function of a low-pass graph convolutional network is formed as

$$Z_{l+1} = \phi\left(\tilde{D}^{\frac{1}{2}} \tilde{A} \tilde{D}^{-\frac{1}{2}} Z_l W^{(l)}\right) \tag{8.20}$$

where Z_l is l-th layer learned representation, and $Z_0 = X$, $W^{(l)}$ is the coefficient matrix learned during the network training.

(2) *Layer-wise Graph Attention Module (LGAT):* As described above, a low-pass graph convolutional filter can be designed by filtering high-frequency basis signals and preserving low-frequency basis signals. However, the low-pass GCN cannot learn layer-wise intermediate information for node clustering. In addition, the available deep clustering methods are usually limited to globalization refinement and neglect the multi-scale spatial-spectral features embedded in different layers [45]. To address these issues and exploit the multi-scale layer-wise locality information, we design the LGAT module.

The features at different layers contain different levels of semantic description to depict the input data, and the final clustering task may require different locality layer-wise information. The conceptual visual interpretation is shown in Fig. 8.2.

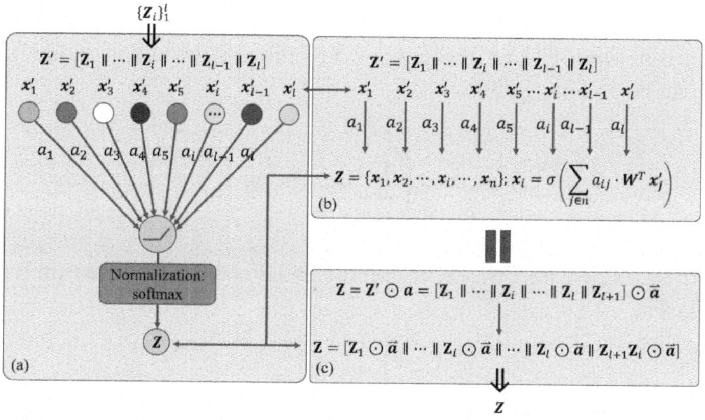

Fig. 8.2 The conceptual illustrations of the proposed layer-wise graph attention (LGAT) module. **a** Is the visual map of LGAT. In **b**, the multi-scale locality layer-wise information is preserved by a concatenation manner, then LGAT is illustrated on a node-wise perspective. In **c**, we express the mechanism of LGAT from a layer-wise perspective. Here, the 'softmax' is the softmax function for normalization. We adopt "↓" to present the input and output of LGAT

As different layers contain different spectral dimensions, we preliminarily preserve the multi-scale locality layer-wise information with a concatenation manner, namely

$$\mathbf{Z}' = \left[\mathbf{Z}_1 \| \cdots \| \mathbf{Z}_i \| \cdots \| \mathbf{Z}_l \| \mathbf{Z}_{l+1}\right] \tag{8.21}$$

where l is the graph convolutional layer, $\mathbf{Z}_i \in \mathbb{R}^{n \times d_i}$ is the ith layer graph features representations with d_i dimension. Particularly, $\mathbf{Z}' = \left\{\boldsymbol{x}_1', \boldsymbol{x}_2', \ldots, \boldsymbol{x}_n'\right\} \in \mathbb{R}^{n \times d_z}$, where $z = \sum_{i=1}^{l} d_i, \boldsymbol{x}_2'$ is the node feature vector aftering concatenation operation.

After that, we employ the graph attention mechanism to extract the concatenated features $\mathbf{Z}'$, the attention coefficient a_{ij} between node i and node j can be formulated as

$$\alpha_{ij} = softmax\left(e_{ij}\right) = \frac{\exp\left(\sigma\left(\boldsymbol{a}^T\left[\boldsymbol{W}\boldsymbol{x}_i' || \boldsymbol{W}\boldsymbol{x}_j'\right]\right)\right)}{\sum_{j \in n} \exp\left(\sigma\left(\boldsymbol{a}^T\left[\boldsymbol{W}\boldsymbol{x}_i' || \boldsymbol{W}\boldsymbol{x}_j'\right]\right)\right)} \tag{8.22}$$

where $\boldsymbol{a}^T$ is the parameter vector, σ is activated function, i.e., *LeakyReLU* $(\cdot)$. Then, the attention-based scale-wise strategy is imposed to Eq. (8.21), and the multi-scale locality layer-wise features are weighted by the attention coefficient, which is as follows

$$\begin{aligned} \mathbf{Z} = \mathbf{Z}' \odot \boldsymbol{a} &= \left[\mathbf{Z}_1 \| \cdots \| \mathbf{Z}_i \| \cdots \| \mathbf{Z}_l \| \mathbf{Z}_{l+1}\right] \odot \vec{\boldsymbol{\alpha}} \\ &= \left[\mathbf{Z}_1 \odot \vec{\boldsymbol{\alpha}} \| \cdots \| \mathbf{Z}_i \odot \vec{\boldsymbol{\alpha}} \| \cdots \| \mathbf{Z}_l \odot \vec{\boldsymbol{\alpha}} \| \mathbf{Z}_{l+1} \odot \vec{\boldsymbol{\alpha}}\right] \end{aligned} \tag{8.23}$$

where $\overrightarrow{\boldsymbol{\alpha}}$ is the attention coefficient vector learned by Eq. (8.22). Referring to the propagation function of GAT, the graph convolutional form for output features of each node can be calculated as

$$\boldsymbol{x}_i = \sigma\left(\sum_{j\in n} \alpha_{ij} \cdot \boldsymbol{W}^T \boldsymbol{x}_j'\right) \tag{8.24}$$

From Eq. (8.23), we note that $\boldsymbol{x}_i$ contains rich layer-wise information for different clustering tasks.

8.3.3 *Graph Convolutional Embedding Decoder*

In our proposed method, we develop decoders, i.e., graph reconstruction decoder and graph convolutional embedding clustering decoder, to decoder the hidden representation **Z** learned by locality preserving low-pass graph convolutional embedding encoder and update the trainable weight of L2GCC.

(1) *Graph Reconstruction Decoder:* To reconstruct both node features and the graph structure, a simple inner product decoder is employed to present the reconstructed adjacency matrix in our proposed method, i.e.,

$$\widehat{\boldsymbol{A}}_{ij} = sigmoid(\boldsymbol{Z}^T\boldsymbol{Z}) \tag{8.25}$$

where $\widehat{\boldsymbol{A}}_{ij}$ is the adjacency matrix of a reconstructed graph, which is flexible and efficient to represent the structure of the reconstructed graph. In this way, we can decoder the content and structure of the original graph. To measure the difference between the reconstructed graph adjacency matrix $\widehat{\boldsymbol{A}}$ and the original graph adjacency matrix $\boldsymbol{A}$, we utilize Kullback–Leibler (KL) Divergence to measure the difference between two distributions. To reduce the influence of individual values in the matrix on the final result, the matrix $\boldsymbol{A}$ and $\widehat{\boldsymbol{A}}$ should be normalized first. Then, the graph reconstruction loss is formulated as

$$L_r = KL\left(\boldsymbol{A}\|\widehat{\boldsymbol{A}}\right) = \sum_i\sum_j \widehat{\boldsymbol{A}}_{ij}\log\frac{\boldsymbol{A}_{ij}}{\widehat{\boldsymbol{A}}_{ij}} \tag{8.26}$$

Graph Convolutional Embedding Self-Training Clustering Decoder: As mentioned above, a graph hidden representation **Z** can be learned by a locality preserving low-pass graph convolutional embedding autoencoder. However, the clustering does not participate in the autoencoder process, and it is naive to separate clustering from embedding learning, as the learned **Z** may not suit for clustering. Moreover, the main challenge for unsupervised graph clustering is the nonexistence of labels to train the network. To overcome the problems, inspired by DEC [42], we develop a self-training strategy, in which the self-training clustering objective employs soft labels to supervise the clustering process and obtain appropriate hidden representations for node clustering, the self-training clustering loss is formulated as

$$L_c = KL(P\|Q) = \sum_i \sum_j p_{ij} \log \frac{p_{ij}}{q_{ij}} \tag{8.27}$$

where $KL(\cdot\|\cdot)$ is Kullback–Leibler divergence; p_{ij} is the target distribution; Q is soft labels distribution. We can adopt q_{ij} to measures the similarity between cluster center μ_j and node embedding z_i of node i, which is measured by a student's t-distribution, namely

$$q_{ij} = \frac{\left(1 + \|z_i - \mu_j\|^2\right)^{-1}}{\sum_{j'} \left(1 + \|z_i - \mu_{j'}\|^2\right)^{-1}} \tag{8.28}$$

we can treat each node as a soft clustering assignment distribution. And, we define the target distribution p_{ij} as

$$p_{ij} = \frac{q_{ij}^2 / \sum_i q_{ij}}{\sum_k \left(q_{ik}^2 / \sum_i q_{ik}\right)} \tag{8.29}$$

where $\sum_i q_{ij}$ is a normalization to normalize each centroid's loss contribution and prevent hidden feature space from being distorted by large clusters.

As described above, a denser Q distribution can be by minimizing L_c. Namely, it makes the distribution of nodes in the class more similar, and the distance between nodes in different classes is longer. In this way, we can transform unsupervised clustering into "semi-supervised" clustering.

8.3.4 Joint Embedding and Optimization

In our method, the L2GCC is optimized by integrating the graph embedded autoencoder and self-training clustering, and then we can define the total objective function as

$$L = L_r + \gamma L_c \tag{8.30}$$

where L_r denotes the reconstruction loss of graph autoencoder in Eq. (8.26), L_c is the self-training clustering loss formulated in Eq. (8.27), $\gamma \geq 0$ is a coefficient weight that is used to controls embedded space degree distorting and set as 0.1 in our method.

We should update three paraments by adopting Gradient Descent in the L2GCC, including the coefficient weight $W^{(l)}$, cluster center μ_j and target distribution p_{ij}.

- **The weights and cluster centers updating.** The gradient of L_c relative to μ_j with the fixed p_{ij} can be expressed as

$$\frac{\partial L_c}{\partial \mu_j} = 2\sum_{i=1}^{N}\left(1 + \|z_i - \mu_j\|^2\right)^{-1}\left(q_{ij} - p_{ij}\right)\left(z_i - \mu_j\right) \tag{8.31}$$

Given a learning rate η, we can update the μ_j by

$$\mu_j = \mu_j - \eta\frac{\partial L_c}{\partial \mu_j} \tag{8.32}$$

Then the coefficient weight $W^{(l)}$ is updated by

$$W^{(l)} = W^{(l)} - \eta\left(\frac{\partial L_r}{\partial W^{(l)}} + \gamma\frac{\partial L_c}{\partial W^{(l)}}\right) \tag{8.33}$$

- **The target distribution updating.** The target distribution p_{ij} can be treated as "ground-truth labels" during the training process and should be updated in every T iteration using all embedded nodes. The m initial centroids $\{\mu_j\}_{j=1}^{m}$ are obtained by conducting k-means clustering on the embedded nodes $\mathbf{Z}$. Then the p_{ij} is updated by Eqs. (8.28) and (8.29), and the label of n_i could be estimated by

$$n_i = \underset{j}{argmax} q_{ij} \tag{8.34}$$

To this end, the L2GCC autoencoder is first trained without self-training clustering to learn a well-presentation $\mathbf{Z}$ as demonstrated in Sect. 8.2. Then, we perform self-training clustering to improve the embedding. The training progress will be stopped as the target distribution p_{ij} change between two consecutive updates is less than a fixed δ or the training epochs T reach the upper set limit. The clustering result is presented by the last optimized soft labels distribution Q. The training process of the proposed L2GCC is shown in Algorithm 1.

Algorithm 1: L2GCC training process.

Input: Graph $\mathcal{G}$; Number of initial centroids m; Number of iterations T; Coefficient weight γ; Distribution threshold δ.

Output: Clustering results Q.

Initialization: $t = 1$; $\boldsymbol{Z}^{(0)} = \boldsymbol{X}$; Initialize the parameters of graph autoencoder;
Calculate graph Laplacian $\tilde{\boldsymbol{L}}_{sym}$ by Eq. (8.17);
$k \leftarrow 1/\lambda_{max}$;
Get the graph hidden presentation $\boldsymbol{Z}$ by performing graph autoencoder according to minimizing Eq. (8.26);
Get m initial clusters centroid μ_j by k-means clustering based on $\boldsymbol{Z}$.
for$t$=1 to $T-1$ do
  The soft labels distribution Q is calculated based on μ and $\boldsymbol{Z}$ by Eq. (8.28);
  The target distribution P is calculated based on Q by Eq. (8.29);
  The self-training clustering loss L_c is calculated by Eq. (8.27);
  if $L_c \leq \delta$ **break**;
  Calculate joint embedding loss L by Eq. (8.30) and update the weight matrices adopting Adam gradient descent;
end
Obtain the final clustering results Q according to Eq. (8.34).

8.3.5 HSI Pre-processing and Graph Construction

HSI contains rich spectral information and a large number of pixels, which not only provides sufficient features for target classification but also brings a huge amount of calculation. In practice, the amount of calculation complexity of graph convolutional neural network (GNN) increases exponentially with the increase of the number of nodes. It is unrealistic to calculate each pixel as a node of the graph. In other words, how to reduce the number of nodes while preserving the hyperspectral spectral-spatial characteristics is a major problem in HSI pre-processing. To construct the graph that can then be clustered by the subsequent network, we introduce an HSI pre-proposing method, which contains two parts, including spatial module and spectral module.

- **Spatial** module: Given an HSI cube $I_B = \{x_1, x_2, \cdots, x_m\}$ with $m = \mathcal{W} \times \mathcal{H}$ pixels and $\mathcal{B}$ spectral bands, where $\mathcal{W}$, $\mathcal{H}$ denote the width and height respectively. Firstly, we adopt unsupervised PCA [46] to reduce the dimensionality of HSI, and the first principal component is preserved to generate a reduced image $I_b = \{x_1^r, x_2^r, \cdots, x_m^r\}$. Then, we utilize the SLIC [47] to assign the pixels into superpixels, the mathematical express of local superpixel can be defined as

$$HSI = \cup_{i=1}^{K} S_i, S_i \cap S_j = \varnothing, i \neq j; i, j = 1,2, \ldots, K \tag{8.35}$$

where $S_i = \{p_{i,1}, \ldots, p_{i,n_i}\}$ denotes a superpixel containing n_i connected pixels, K is the total number of superpixels.

- **Spectral** module: As shown above, we can partition the HSI into superpixels by the spatial module. However, the dimensionality reduction loses the spectral information of HSI, namely the superpixel does not fully extract the spectral features of HSI. To solve the problem, we extract the spectral value of each pixel from the original HSI directly, and then we caculate the mean spectral value of pixels in superpixel as the superpixel's spectral features. We can express the spectral feature vector of the pixel at the spatial location $p_0 = (x, y)$ as

$$\boldsymbol{x}(\boldsymbol{p}_0) = (X_1(\boldsymbol{p}_0), X_2(\boldsymbol{p}_0), \cdots, X_{\mathcal{B}}(\boldsymbol{p}_0)) \tag{8.36}$$

where $X_i(\boldsymbol{p}_0)$ denotes the spectral value of the pixel in ith channel at the spatial location $\boldsymbol{p}_0$.

Then, the mathematical expression of the HSI feature $\boldsymbol{X}$ is

$$\begin{aligned} \boldsymbol{X} &= [\boldsymbol{X}_1, \boldsymbol{X}_2, \ldots \boldsymbol{X}_S]^T \\ &= \left[\frac{1}{\tau_1}\sum_{k=1}^{N_1}\boldsymbol{x}_k^1, \frac{1}{\tau_2}\sum_{k=1}^{N_2}\boldsymbol{x}_k^2, \ldots, \frac{1}{\tau_K}\sum_{k=1}^{N_S}\boldsymbol{x}_k^S\right]^T \end{aligned} \tag{8.37}$$

where $\boldsymbol{X}_i$ is the ith superpixel feature vector, and τ_i denotes the number of pixels contained in superpixel, $\boldsymbol{x}_k^i$ is the spectral feature vector of the pixel shown as Eq. (8.36).

- **Graph construction:** In the proposed method, we take the superpixels as the graph node, the adjacency matrix $\boldsymbol{A}$, defining the relationships between vertexes, can be generally computed as a radial basis function (RBF)

$$\boldsymbol{A}_{ij} = \begin{cases} e^{-\rho \boldsymbol{X}_i - \boldsymbol{X}_j^2}, & \text{if } \boldsymbol{X}_i \in \mathcal{N}_t(\boldsymbol{X}_j) \text{ or } \boldsymbol{X}_j \in \mathcal{N}_t(\boldsymbol{X}_i) \\ 0, & \text{otherwise} \end{cases} \tag{8.38}$$

where $\rho = 0.2$ is an empirical set to control the RBF's width, $\boldsymbol{X}_i$ denote nodes features of nodes i,$\mathcal{N}_t(\boldsymbol{X}_i)$ is the t-hop neighbors of $\boldsymbol{X}_i$.

8.3.6 Computational Complexity

Let n be the number of graph nodes, d be the input node spectral dimension, d_i be the ith layer node spectral dimension, $|\mathcal{E}|$ be the number of nonzero entries of the adjacency matrix A, c be the number of clusters, l be the number of graphs embedded auto-encoder layers. The computational complexity of low-pass GCN convolution is calculated as $O\left[(n + |\mathcal{E}|)\left(dd_1 + \sum_{i=2}^{l-1} d_i d_{i-1}\right)\right]$. For layer-wise GAT, the computational complexity is $O\left[|\mathcal{E}|\left(\sum_{i=1}^{l-1} d_i\right)d_l\right]$. The computational complexity of graph auto-encoder is $O\left[(n+|\mathcal{E}|)\left(dd_1+\sum_{i=2}^{l-1} d_i d_{i-1}\right)+|\mathcal{E}|\left(\sum_{i=1}^{l-1} d_i\right) d_l\right]$. For Eq. (8.29), the computational complexity is $O[(nc+n\log n)+(n^2+n\log n)]$. Therefore, the computational complexity of performing clustering on the graph in each iteration is $O[(n+|\mathcal{E}|)\left(dd_1+\sum_{i=2}^{l-1} d_i d_{i-1}\right)+|\mathcal{E}|\left(\sum_{i=1}^{l-1} d_i\right) d_l+(nc+n\ \log\ n)+$ (n2 + n log n)]. If our model iterates T times, we express the overall computational complexity of LGCC is $O[[(n+|\mathcal{E}|)\left(dd_1+\sum_{i=2}^{l-1} d_i d_{i-1}\right)+|\mathcal{E}|\left(\sum_{i=1}^{l-1} d_i\right) d_l+$ (nc + n log n) + (n2 + n log n)]T]. From the expression formula of computational complexity, we can greatly reduce the computational complexity by reducing the number of nodes, which verifies the effectiveness of the HSI pre-processing.

8.4 Proofs

Theorem 1. *The smoother signals can be obtained by preserving low-frequency basis signals and filtering high-frequency basis signals.*

Proof. Given the graph signal (node features) $\boldsymbol{X}$, considering $\boldsymbol{L}=\boldsymbol{U}\boldsymbol{\Lambda}\boldsymbol{U}^{-1}$ and referring to Eq. (8.10), we can express the eigenvector $\boldsymbol{u}_{\mathrm{q}}$ smoothness as

$$R(\boldsymbol{L},\boldsymbol{u}_q)=\frac{\boldsymbol{u}_q{}^T\boldsymbol{L}\boldsymbol{u}_q}{\boldsymbol{u}_q{}^T\boldsymbol{u}_q}=\lambda_q \tag{8.39}$$

where $\boldsymbol{u}_q(i)$ is the $\boldsymbol{u}_q$'s i-th element. Equation (8.39) shows that smoother eigenvectors have smaller eigenvalues. Then we can decompose $\boldsymbol{X}$ into a linear combination, i.e.,

$$\boldsymbol{X}=\boldsymbol{U}\boldsymbol{c}=\sum_{q=1}^{n} c_q\boldsymbol{u}_q \tag{8.40}$$

where c_q denotes the coefficient of q-th eigenvector $\boldsymbol{u}_q$. Thus the smoothness of $\boldsymbol{X}$ can be calculated based on Eqs. (8.10), (8.39), and (8.40)

$$R(\boldsymbol{L},\boldsymbol{X})=\frac{\boldsymbol{X}^T\boldsymbol{L}\boldsymbol{X}}{\boldsymbol{X}^T\boldsymbol{X}}=\frac{\sum_{c=1}^{n} c_q^2\lambda_c}{\sum_{c=1}^{n} c_q^2} \tag{8.41}$$

Therefore, the Theorem 1 is proved.

Theorem 2. *Given a graph filter* $\boldsymbol{G}$, *if its frequency response function* $p(\lambda)$ *is nonnegative and nonincreasing for all* λ_i, *then given any signal* $\boldsymbol{X}$ *and the filtered* $\widetilde{\boldsymbol{X}}=\boldsymbol{G}\boldsymbol{X}$, *we always have*

$$R(\widetilde{X}, X) \leq R(L, X) \tag{8.42}$$

Namely

$$\frac{\sum_{i=1}^{n} b_i\lambda_i}{\sum_{i=1}^{n} b_i} \leq \frac{\sum_{i=1}^{n} c_i\lambda_i}{\sum_{i=1}^{n} c_i} \tag{8.43}$$

Proof. We adopt induction to prove the Theorem 2, namely

$$\begin{aligned} R_b^i &= \frac{\sum_{i=1}^{n} b_i\lambda_i}{\sum_{i=1}^{n} b_i} \\ R_c^i &= \frac{\sum_{i=1}^{n} c_i\lambda_i}{\sum_{i=1}^{n} c_i} \end{aligned} \tag{8.44}$$

If, $\lambda_1 \leq \cdots \lambda_n$,and $\frac{c_1}{b_1} \leq \cdots \leq \frac{c_n}{b_n}$, we can easily prove that $R_b^2 \leq R_c^2$.Now, suppose that $n = l - 1$, and $R_b^{l-1} \leq R_c^{l-1}$, then we have R_b^l

$$\begin{aligned} \frac{\sum_{i=1}^{l} b_i\lambda_i}{\sum_{i=1}^{l} b_i} &= \frac{\sum_{i=1}^{l-1} b_i\lambda_i + b_l\lambda_l}{\sum_{i=1}^{l-1} b_i + b_l} \\ &= \frac{\left(\sum_{i=1}^{l-1} b_i\right)\lambda_b^{(l-1)} + b_l\lambda_l}{\sum_{i=1}^{l-1} b_i + b_l} \\ &\leq \frac{\left(\sum_{i=1}^{l-1} b_i\right)\lambda_c^{(l-1)} + b\lambda_l}{\sum_{i=1}^{l-1} b_i + b_l} \end{aligned} \tag{8.45}$$

Since, $R_b^{l-1} \leq R_c^{l-1}$, and $\frac{\sum_{i=1}^{l-1} c_i}{\sum_{i=1}^{l-1} b_i} \leq \frac{c_l}{b_l}$, we have

$$\begin{aligned} \frac{\left(\sum_{i=1}^{l-1} b_i\right)\lambda_c^{(l-1)} + b\lambda_l}{\sum_{i=1}^{l-1} b_i + b_l} &\leq \frac{\left(\sum_{i=1}^{l-1} c_i\right)\lambda_c^{(l-1)} + c\lambda_l}{\sum_{i=1}^{l-1} c_i + c_l} \\ &= \frac{\sum_{i=1}^{l} c_i\lambda_i}{\sum_{i=1}^{l} c_i} \end{aligned} \tag{8.46}$$

Hence, the Theorem 2 holds for all n.

Table 8.2 The details of the datasets used for the evaluation of the investigated algorithms

No.	Indian pines		Salinas		UH2013	
	Landcover name	Pixels	Landcover name	Pixels	Landcover name	Pixels
1	Asphalt	46	Weed 1	2009	Healthy-grass	1251
2	Corn-notill	1428	Weed 2	3726	Stressed-grass	1254
3	Corn-mintill	830	Fallow	1976	Synthetic-grass	697
4	Corn	237	Fallow-plow	1394	Tress	1244
5	Grass-pasture	483	Fallow-smooth	2678	Soil	1242
6	Grass-trees	730	Stubble	3959	Water	325
7	Grass-pasture-mowed	28	Celery	3579	Residential	1268
8	Hay-windrowed	478	Grapes untrained	11,271	Commercial	1244
9	Oats	20	Soil	6203	Road	1252
10	Soybean-notill	972	Corn	3278	Highway	1227
11	Soybean-mintill	2455	Lettuce-4 week	1068	Railway	1235
12	Soybean-clean	593	Lettuce-5 week	1927	Parking-Lot 1	1233
13	Wheat	205	Lettuce-6 week	916	Parking-Lot 2	469
14	Woods	1265	Lettuce-7 week	1070	Tennis-court	428
15	Buildings-grass-trees-drives	386	Vineyard-untrained	7268	Running-track	660
16	Stone-steel-towers	93	Vineyard-trellis	1807		

8.5 Experiments

8.5.1 Datasets

In the experiments, we employ three widely-used HSI datasets to test the performances of the investigated methods, including Indian Pines (IP), Salinas, and University of Houston 2013 (UH2013) datasets. The Indian Pines dataset contains 200 spectral bands with an image size of 145 × 145 and 16 landcover classes. The Salinas dataset is composed of 204 spectral channels with a size of 610 × 340 and nine land-cover classes. For the Houston 2013 dataset, the data size is 349 × 1905 with 144 channels. The image is labeled 15 categories. The details of the three datasets used in the experiment are illustrated in Table 8.2.

8.5.2 Baselines and Evaluation Metrics

To verify the clustering accuracy and real-time performances of the proposed L2GCC, we consider 10 baselines for comparison, including *k*-means [15], FCM [17], FCM-S1 [22], spectral clustering (SC) [16], SSSC [24], S^5C [25], DS^3C-Net [33], AE

Table 8.3 Optimal values utilised in L2GCC

Dataset	S	L	T	η	c	γ	δ (%)
IP	1000	3	200	0.001	16	0.1	0.1
Salinas	7000	3	200	0.001	16	0.1	0.1
UH 2013	14,000	3	200	0.001	15	0.1	0.1

[48] + k-means, DEC [30], and CC [35]. Notably, k-means, FCM, FCM-S1, and SC are traditional clustering methods designed for natural images, SC and CC are spectral clustering method and contrastive clustering method respectively. SSSC and S^5C belong to sparse subspace clustering methods. Furthermore, DS^3C-Net, AE + k-means, DEC, and CC are deep clustering methods and are developed for large-scale HSI clustering. The hyperparameters of the aforementioned methods are the same as their original papers.

In our experiments, five metrics, including overall accuracy (OA), per-class accuracy (PA), kappa coefficient (κ), normalized mutual information (NMI), and adjusted rand index (ARI), are introduced for all the investigated methods to estimate their performances.

8.5.3 *Experimental Settings*

In our method, there are mainly six parameters, including the number of superpixels S (graph node), number of low-pass graph convolutional layers L, learning rate η, number of epochs T, number of initial centroids c, coefficient weight γ, and distribution threshold δ, should be set, and their optimal values utilized in the proposed method are shwn in Table 8.3. It should be noted that the setting of δ is very small and will not be triggered under normal circumstances, except for the discrete case of clustering. The c initial centroids $\{\mu_j\}_{j=1}^{c}$ are obtained by conducting k-means clustering, and its value is equal to the number of land-covers contained in the HSI. The dimensions of the three hidden layers are 168, 84, and 42, respectively.

The experiments in the paper are implemented on the Intel i9-10900 K processor with 3.70G of DDR4 RAM. Regarding the graphic processing unit (GPU), the NVIDIA GeForce GTX 1080Ti with 11G of DDR4 RAM is used. To minimization random influence, we execute all experiments ten times and report the average clustering performances with Pytorch-Geometric as a back-end.

8.5.4 *Experimental Performances Comparison*

In this section, the quantitative clustering results on the IP, Salinas, and UH2013 datasets are reported in detail in Table 8.4, 8.5, and 8.6. The best performances and the second-best performances are marked in bold and italic, respectively. The corresponding visual clustering results are demonstrated in Figs. 8.3, 8.4, and 8.5.

Quantitative results for the IP dataset: From the quantitative clustering results in Table 8.4, the following conclusions can be easily summarized. First, L2GCC achieves the best results among all key indicators in the experimental results, which shows that our model has a good filtering ability for interference noise. Specifically, compared with the second-best results, the results of our proposed model improve by 18.37, 16.37, 8.72, and 12.56% in OA, κ, NMI, and ARI, respectively. Second, compared with classical methods, i.e., k-means, FCM, FCM-S1, scalable spectral/ subspace clustering methods can achieve superior performances since their mechanism design can help to feature extraction. Specifically, SC, SSSC, and S^{5}C have higher clustering accuracies, and the OAs of the three methods are improved by 3.42, 0.17, and 2.61%, respectively, compared with k-means. The deep clustering algorithms (i.e., DS^{3}C-Net, AE + k-means, DEC, CC, and L2GCC) are significantly better than shallow methods (i.e., k-means, FCM, FCM-S1, SC), mainly because deep learning can extract high-level features for clustering. Specifically, compared with the best results in shallow methods, the clustering accuracies of deep clustering algorithms in terms of the OA are increased by 3.77, 3.05, 0.18, and 13.16%. It also should be noted that the graph-based clustering method DS^{3}C-Net obtains competitive results among deep clustering methods since it can extract the complex nonlinear similarities among pixels by converting high-dimensional spectral features into a subspace.

Quantitative results for the Salinas dataset: From Table 8.5, the following conclusions can be obtained. First, k-means achieves the best results in classical methods, which shows that k-means can successfully identify more categories than the other two methods. Specifically, OA of k-means is 38.11%, which is improved by 0.54 and 2.22% compared with the value of FCM (37.57%) and FCM-S1 (35.89%). Second, for the spectral/ subspace clustering methods, SSSC achieves more effective performance, its OA (52.43%), κ (50.69%), NMI (64.23%), and ARI (46.76%) are improved by 0.75, 0.34, 0.96, and 3.17% compared with these values of S^{5}C. Furthermore, the performances of SSSC and S^{5}C are better than the classical methods. However, SC failed to handle the clustering because of out-of-memory, which shows that SC needs to be improved in the clustering adaptability of large datasets. Third, for the quantitative results obtained by deep and shallow clustering methods, different from our consideration, deep clustering methods do not achieve the expected clustering advantages, which illustrates that deep clustering methods are not always superior to shallow clustering methods. However, DS^{3}C-Net obtains more competitive performance and improves OA 1.05, 4.16, and 2.77% in contrast to the value of AE + k-means, DEC, and CC, respectively. Fourth, the proposed L2GCC achieves the best clustering results by improving the OA, κ, NMI, and ARI by 13.74, 12.09, 5.84, and 10.04% compared with this value of the second-best method, which signifies the effectiveness of the proposed method.

Table 8.4 Quantitative experimental performances of investigated methods on IP, bold denotes the best performances

No	*k*-means	FCM	FCM-S1	SC	SSSC	S^5C	DS^3C-Net	AE + *k*-means	DEC	CC	L2GCC
1	0.0594	0.0567	0.0634	0.0000	0.0236	0.1125	0.1142	0.0870	0.0000	0.0870	**0.2543**
2	0.4563	0.5226	0.4426	0.0273	0.1836	0.2537	0.2709	0.1408	0.1877	0.2955	**0.6787**
3	0.1819	0.2118	0.1781	0.0048	0.3216	0.3576	**0.6985**	0.1940	0.2892	0.5795	**0.6028**
4	0.1237	0.1058	0.2109	0.0000	0.3018	0.1531	0.8861	0.1730	0.2405	0.8861	**0.9017**
5	0.3447	0.2946	0.3393	0.0166	0.6825	0.6279	**0.6534**	0.5217	0.0000	0.6584	**0.8236**
6	0.6901	0.7645	0.8008	0.0000	0.4913	0.5524	0.9795	0.9329	0.9726	0.9795	**1.0000**
7	0.0000	0.1043	0.0000	0.0000	0.6127	**0.7019**	0.0000	0.0000	0.0000	0.0000	0.0000
8	0.6239	0.8922	0.5706	**1.0000**	0.5736	0.5124	0.9137	0.7134	0.6464	**1.0000**	**1.0000**
9	0.0000	**0.6400**	0.0215	0.0000	0.0000	0.1329	0.0000	0.0000	0.0000	0.0000	0.3285
10	0.5654	0.2528	0.5268	0.0031	0.3162	0.2618	0.5327	0.5021	0.3693	0.4095	**0.7021**
11	0.4733	0.5141	0.5966	**0.9609**	0.3769	0.2235	0.2957	0.3752	0.3202	0.2957	0.6833
12	0.2592	0.2788	0.2553	0.0017	0.1823	0.0000	0.6572	0.1551	0.2749	**0.7437**	0.4564
13	0.3276	0.4289	0.2956	0.1463	0.6571	0.7531	**1.0000**	0.5024	0.3610	**1.0000**	**1.0000**
14	0.8559	0.8006	0.9173	**1.0000**	0.4839	0.4629	0.6854	0.7605	0.7763	0.6854	0.8346
15	0.2007	0.2869	0.2584	0.0000	0.2649	0.0000	**1.0000**	0.6995	0.7021	**1.0000**	**1.0000**
16	0.0000	0.0632	0.0000	0.7419	0.68310	0.7726	0.1986	0.5591	0.6022	0.0000	**0.8235**
OA	0.3811	0.3757	0.3589	0.4153	0.3828	0.4072	*0.5738*	0.4458	0.4171	0.5514	**0.7351**
κ	0.3572	0.2975	0.2782	0.2794	0.3326	0.3518	*0.5364*	0.3847	0.3591	0.5192	**0.6829**
NMI	0.4386	0.4380	0.4217	0.5041	0.4062	0.4317	*0.6928*	0.4527	0.4787	0.6644	**0.7516**
ARI	0.2195	0.2295	0.2070	0.2455	0.2138	0.2296	*0.4350*	0.2661	0.2968	0.4070	**0.5326**

Table 8.5 Quantitative experimental performances of investigated methods on Salinas, bold denotes the best performances

No	*k*-means	FCM	FCM-S1	SC	SSSC	S^5C	DS^3C-Net	AE + *k*-means	DEC	CC	L2GCC
1	0.9985	0.4179	0.4860	OOM	0.0416	0.0000	0.8761	0.0000	0.9000	**1.0000**	**1.0000**
2	0.5698	0.5524	0.5637	OOM	0.9875	0.9628	0.9473	**1.0000**	0.7933	0.8969	0.9826
3	0.9741	0.8501	**1.0000**	OOM	0.0154	0.0434	**1.0000**	0.5501	0.0000	**1.0000**	0.9234
4	0.9865	0.0000	0.9537	OOM	0.8131	0.8512	0.6502	0.9340	0.5079	0.0000	0.8762
5	**0.7856**	0.5540	0.7365	OOM	0.8723	0.7526	**0.8490**	0.7367	0.7412	0.5310	0.7731
6	0.9952	0.9990	0.9704	OOM	0.9358	0.9832	0.7614	0.8406	0.8366	0.6603	**1.0000**
7	0.4810	0.0000	0.0003	OOM	0.5412	0.1126	0.5236	0.3328	0.6943	0.7499	0.7806
8	0.7309	0.7772	0.8412	OOM	0.4226	0.4917	0.4608	0.5008	**0.8565**	0.3615	0.8219
9	0.9750	0.6969	0.9602	OOM	0.5638	0.6826	0.6602	**0.9997**	0.9994	0.7201	0.9239
10	0.6169	0.5507	0.5429	OOM	0.1143	0.3158	0.8133	0.5863	0.3350	0.9451	**0.9137**
11	0.3370	0.0000	0.3013	OOM	0.6915	0.6213	**0.8315**	0.0000	0.0000	**1.0000**	0.7129
12	0.0273	0.0047	0.0000	OOM	0.4857	0.5184	**1.0000**	0.9460	0.8293	**0.9984**	0.9032
13	0.3548	0.8134	0.4613	OOM	0.0000	**0.9321**	0.0000	0.0000	0.0000	0.0000	0.6513
14	0.5584	0.6512	0.7469	OOM	0.1826	**0.8725**	0.0000	0.9645	0.0000	0.0000	0.6228
15	0.4994	0.4279	0.5064	OOM	0.4682	0.3948	0.5687	0.6237	0.0579	0.5687	**0.8027**
16	0.0000	0.8321	0.8433	OOM	0.9265	0.6617	**0.9182**	0.9917	0.9092	**1.0000**	0.8245
OA	*0.6799*	0.5673	0.6318	OOM	0.5243	0.5168	0.6673	0.6568	0.6257	0.6396	**0.8315**
κ	*0.6572*	0.5431	0.6202	OOM	0.5069	0.5035	0.6326	0.6206	0.5796	0.6123	**0.8226**
NMI	0.7356	0.6859	0.7089	OOM	0.6423	0.6327	0.7785	0.7720	0.7705	*0.7806*	**0.8569**
ARI	0.5466	0.4693	0.5119	OOM	0.4676	0.4359	*0.5531*	0.5548	0.5824	0.5330	**0.6926**

"OOM" denotes out of memory

Table 8.6 Quantitative experimental performances of investigated methods on UH2013, bold denotes the best performances

No	*k*-means	FCM	FCM-S1	SC	SSSC	S^5C	DS^3C-Net	AE + *k*-means	DEC	CC	LGCC
1	0.7926	0.4932	0.4762	OOM	0.5238	0.4813	**0.9235**	0.8129	0.8185	0.7506	**0.9017**
2	0.1436	0.2831	0.3017	OOM	0.3671	0.2516	0.8702	0.0973	**0.7783**	0.6938	0.4826
3	0.7638	0.0000	0.0000	OOM	0.0000	0.0000	0.0000	0.9842	0.4390	0.9900	**0.9999**
4	0.7024	0.3834	0.3921	OOM	0.4229	0.4420	0.8135	0.7436	0.0000	0.6624	**0.8731**
5	0.0000	0.6932	0.6273	OOM	0.2783	0.0000	0.7761	0.8285	**0.9815**	0.7198	0.7531
6	0.0000	0.0185	0.0372	OOM	0.1541	0.1266	0.2118	**0.7600**	0.0000	0.0000	0.6128
7	0.2416	0.0789	0.0564	OOM	0.0000	0.0510	0.3970	**0.8028**	0.0000	0.6088	0.6529
8	0.0000	0.0949	0.0629	OOM	0.2576	0.1932	0.6290	0.2291	0.2098	0.3778	**0.5326**
9	0.2671	0.1070	0.1324	OOM	0.0962	0.1224	0.4702	0.0000	**0.4744**	0.3011	0.0932
10	0.2935	0.2347	0.2179	OOM	0.2967	0.3125	0.5213	0.2771	**0.6544**	0.3325	0.5932
11	0.2613	**0.5417**	0.4386	OOM	0.3629	0.2718	0.3926	0.2777	0.0089	0.2874	0.4217
12	0.7132	0.8021	0.8211	OOM	0.7941	0.8530	0.6211	0.7875	0.5929	0.4039	**0.8871**
13	0.0000	0.0000	0.000	OOM	0.0000	0.0000	0.4425	0.0000	**0.6482**	0.6077	0.6168
14	**1.0000**	0.0000	0.0000	OOM	0.0000	0.0000	0.0000	0.9907	**1.0000**	0.0000	0.5132
15	0.3818	0.0000	0.0427	OOM	0.0000	0.0000	0.5233	0.5727	0.6258	0.9924	**1.0000**
OA	0.3976	0.3070	0.3182	OOM	0.3352	0.3167	*0.5473*	0.5180	0.4704	0.5348	**0.6412**
κ	0.3641	0.2456	0.2573	OOM	0.3219	0.3221	*0.5126*	0.4783	0.4278	0.5008	**0.6315**
NMI	0.5217	0.3623	0.3716	OOM	0.4153	0.4078	*0.6094*	0.5871	0.5660	0.5949	**0.6931**
ARI	0.2638	0.1488	0.1534	OOM	0.2035	0.1726	*0.4103*	0.3613	0.3330	0.3977	**0.5228**

"OOM" denotes out of memory

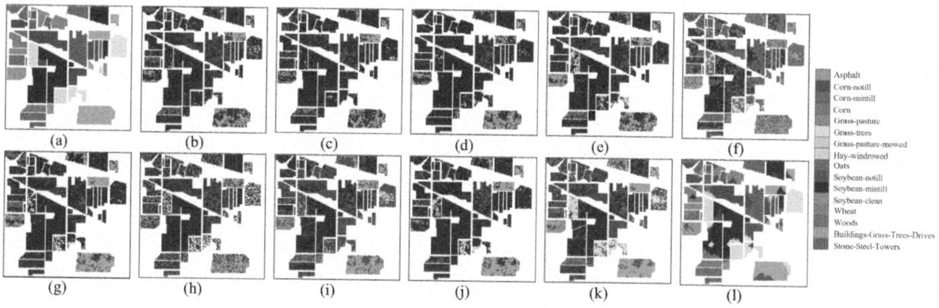

Fig. 8.3 The visual clustering map of investigated methods on IP dataset. **a** Ground-truth map; **b** k-means (OA = 38.11%); **c** FCM (OA = 37.57%); **d** FCM-S1 (OA = 35.89%); **e** SC (OA = 41.53%); **f** SSSC (OA = 38.28%); **g** S^5C (OA = 40.72%); **h** DS^3C-Net (OA = 57.38%); **i** AE + k-means (OA = 44.58%); **j** DEC (OA = 41.71%); **k** CC (OA = 55.14%); **l** L2GCC (OA = 73.51%)

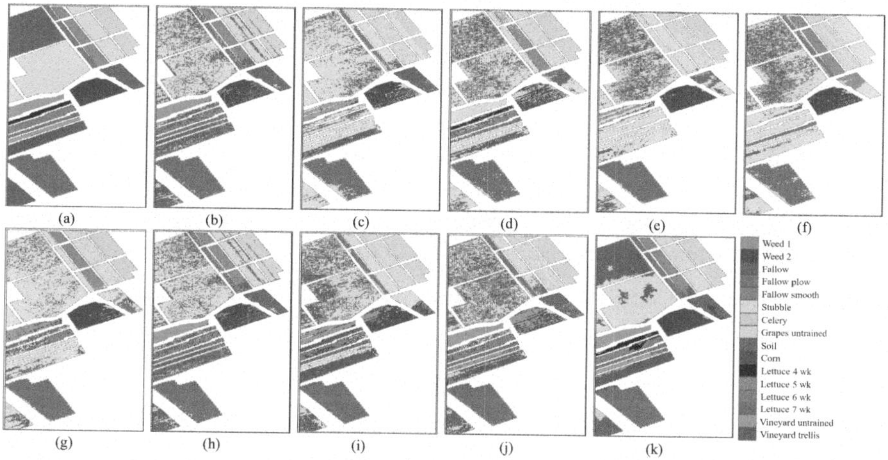

Fig. 8.4 The visual clustering map of investigated methods on Salinas dataset. **a** Ground-truth map; **b** k-means (OA = 67.99%); **c** FCM (OA = 56.73%); **d** FCM-S1 (OA = 63.18%); **e** SSSC (OA = 52.43%); **f** S^5C (OA = 51.68%); **g** DS^3C-Net (OA = 66.73%); **h** AE + k-means (OA = 65.68%); **i** DEC (OA = 62.57%); **j** CC (OA = 63.96%); **k** L2GCC (OA = 83.15%)

Quantitative results for the UH2013 dataset: From the quantitative clustering results in Table 8.6, the following conclusions can be easily summarized. First, compared with the clustering results of IP and Salinas datasets, the clustering results of all investigated methods in UH2013 are greatly decreased. The main reason is that the UH2013 is larger and contains more pixels and complex interrelationships, which poses higher challenges to the investigated algorithms, and SC is suffering from out-of-memory. Second, different from the results in Salinas, deep clustering methods achieve superior results than shallow methods, which indicates the good adaptability to large-scale dataset clustering of deep clustering methods by learning the high-level semantic features to measure interrelation between raw data. For example,

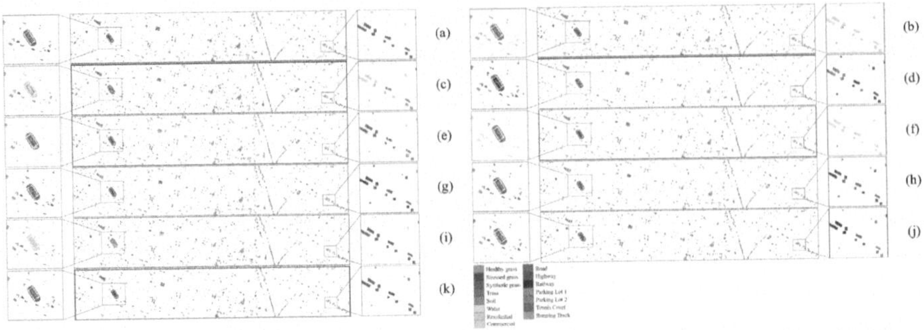

Fig. 8.5 The visual clustering map of investigated methods on UH2013 dataset. **a** Ground-truth map; **b** k-means (OA = 39.76%); **c** FCM (OA = 30.70%); **d** FCM-S1 (OA = 31.82%); **e** SSSC (OA = 33.52%); **f** S^5C (OA = 31.67%); **g** DS^3C-Net (OA = 54.73%); **h** AE + k-means (OA = 51.80%); **i** DEC (OA = 47.04%); **j** CC (OA = 53.48%); **k** L2GCC (OA = 64.12%)

OAs achieved by AE + k-means and CC are 51.80 and 53.48% respectively, which improves the best performance in shallow methods by 12.04 and 13.72%. Furthermore, compared with the k-means, the deep version of AE + k-means obtains a considerable improvement in terms of clustering accuracy. Third, L2GCC still achieves the best quantitative results, which further validates the superiority of our method design. Generally speaking, the proposed L2GCC method not only enables filter image noise but also has good adaptability to different datasets.

Visual Clustering Results: The visual clustering map of investigated methods on IP, Salinas, and UH2013 are shown in Figs. 8.3, 8.4, and 8.5. From the results, we can easily observe that our LGCC achieves the least misclassification clustering maps, which are noticeably closer to the ground truth. In contrast, the comparative methods yield poorer clustering maps with obvious salt and pepper noises. This is because the low-pass graph filter adopted in the L2GCC can preserve the low-frequency signal, which helps for clustering. The layer-wise local features learned by LGAT provide rich features for clustering. Furthermore, high-level features learned by L2GCC contribute to achieving a more accurate clustering map.

8.5.5 *t-SNE Data Distributions Visualization*

In this experiment, the t-distributed stochastic neighbor embedding (t-SNE) [44] is adopted to visualize the change of data distributions of IP, Salinas, and UH2013 with L2GCC training. The 2D visualization of the L2GCC on three datasets at the epoch 0 and 200 are shown in Fig. 8.6. From the results, we can conclude that our L2GCC can well distinguish different types of nodes with the increasing training iteration. Specifically, the small interclass dissimilarity and the large intraclass variations existed in the original three datasets (first row). As network training, L2GCC

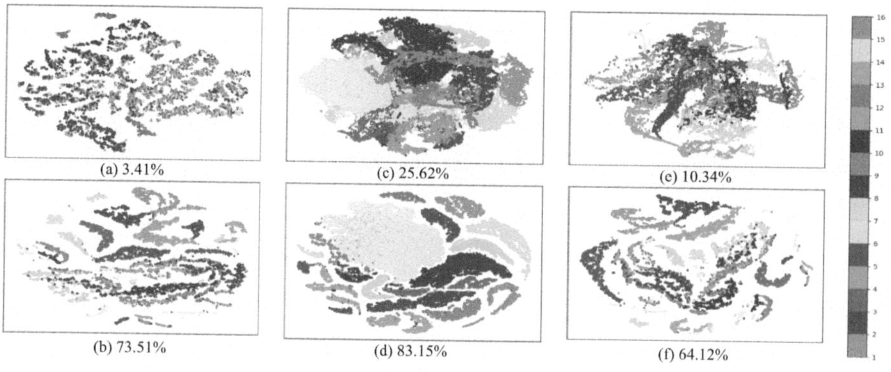

Fig. 8.6 2D visualization of the L2GCC on the IP (first column), Salinas (second column), and UH2013 (third column) at the epoch 0 and 200. Different colours in the maps denote different landcovers

can learn the most discriminative feature representations, which are both intraclass compact and interclass dispense, and contribute to higher clustering accuracy.

8.5.6 Parameter Study

In this section, we investigate the influence of the number of superpixels S (graph node), number of low-pass graph convolutional layers L, learning rate η, and number of epochs T on the performance of L2GCC. We divide into four parameters into two groups, and a grid search strategy is utilized to find the optimal setting. The experimental setups are consistent with Sect. 8.5.3, and the clustering accuracy OA is adopted to report the performance of our L2GCC method with different parameter settings. We show the parament-sensitive results of two group parameters on IP, Salians, and UH2013 in Figs. 8.7 and 8.8.

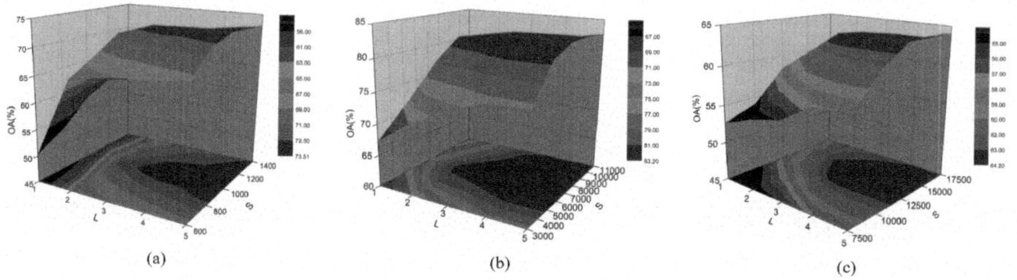

Fig. 8.7 3D visualization of the LGCC with different S and L on the Indian Pines (**a**), Salinas (**b**), and UH2013 (**c**) datasets

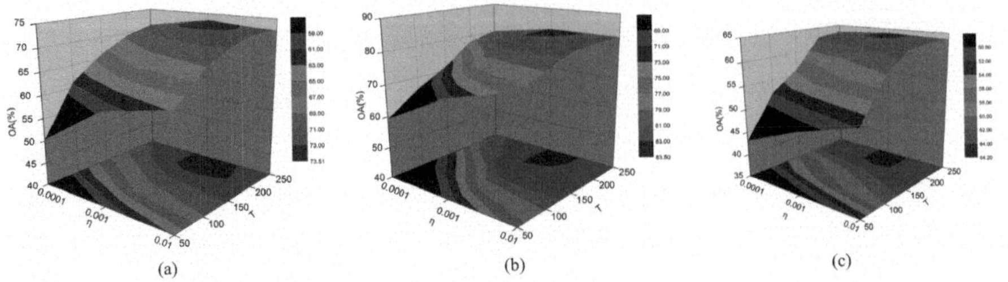

Fig. 8.8 3D visualization of LGCC with different T and η on the Indian Pines (**a**), Salinas (**b**), and UH2013 (**c**) datasets

***S* and *L* effect analysis**: The parament-sensitive results of S and L on L2GCC are shown in Fig. 8.7. In the experiment, S is varied from 600 to 1400 with an interval of 2500 on the IP dataset, from 3000 to 11,000 with an interval of 2000 on the Salinas dataset, and from 7500 to 17,500 on UH2013 dataset, the L is set from 1to 5. Furthermore, we fix T and η as illustrated in Table 8.3. In the results, the OAs on three datasets are being improved as increasing of S. However, the clustering accuracy does not increase after reaching S a certain threshold. The bigger S means smaller superpixels, thus we can preserve more local spatial features of HSI for clustering. However, bigger S leads to more graph node, too many graph nodes will bring challenges to network feature extraction and increase the computational complexity. Therefore, the S is as 1000, 7000, and 14,000 for IP, Salinas, and UH2013, respectively. In terms of L, a bigger L can learn the high-level features of graphs, effectively. However, more weight parameters and more time are required to train the network. In our L2GCC, we can observe that the optimal L is 3 on three datasets.

***T* and *η* effect analysis**: In the experiment, we vary T from 50 to 250 with an interval of 50, and η is set as 0.0001, 0.001, and 0.01. Furthermore, S, L, m, and δ are fixed as in Table 8.3. The parament-sensitive results of T and η on L2GCC are shown in Fig. 8.8. We can train the paraments in the network faster with a large η. However, we usually cannot obtain an optimal solution, and the final results are unstable. Therefore, the η is set as 0.001 in our L2GCC method. It also should be noted that T is closely related to η, and a bigger T means needing more time to train the network. From the results, we can observe that the optimal results occur when T is 200.

8.5.7 *Ablation Study*

In this experiment, we assess the role of pixel to region transformation module (PRT), Low-pass GCN, and LGAT in L2GCC for boosting the HSI clustering on IP, Salinas, and UH2013 datasets. The experimental setup is kept identical to Sect. 8.5.3. The OA is adopted to report the clustering accuracy, and the results are shown in Table 8.7. For simplicity, we utilize "L2GCC-v_1" "L2GCC-v_7" to represent the reduced methods. From the results, we know that L2GCC achieves the best performances. Concretely, the clustering accuracies of LGCC-v_2 are better than the values of L2GCC-v_1, and L2GCC-v_7 performs better than L2GCC-v_3, which indicates that PRT can improve the clustering accuracy. Compared with LGCC-v_1, L2GCC-v_3 achieves a superior result, and LGCC-v_5 outperforms LGGCC-v_4, which verifies Low-pass GCN plays an important role in the proposed methods. In the same way, we can observe that the OA of L2GCC-v_3 is higher than L2GCC-v_1, and L2GCC-v_6 achieves a superior result to that of L2GCC-v_2, which shows the superiority of LGAT. Therefore, PRT, Low-pass GCN, and LGAT contribute to improving the clustering accuracy in our L2GCC.

8.6 Conclusions

In this paper, a novel self-supervised L2GCC framework is developed to boost the clustering performance for large-scale HSI datasets. Specifically, a region transformation module is designed to transform the local spatial-spectral features of the HSI and reduce the computational complexity. The locality preserving low-pass graph convolutional embedding autoencoder is developed to efficiently learn the high-level hidden representations at the superpixel level, in which the low-pass GCN and layer-wise GAT are designed to extract the smoother features and locality features, respectively. Furthermore, the Kullback–Leibler (KL) Divergence is adopted to evaluate the graph reconstruction loss. We perform the self-training clustering with soft labels generated from the graph embedding. The L2GCC algorithm is jointly characterized by locality preserving low-pass graph convolutional embedding autoencoder and self-training clustering, and both components are mutually beneficial. Compared with state-of-the-art clustering methods, L2GCC shows a remarkable model design and superiority, the clustering accuracy OAs are 73.51, 83.15, and 64.12% on IP, Salinas, and UH2013, respectively.

Compared with the supervised and semi-supervised methods, the unsupervised methods are needed to be improved in terms of clustering accuracy, especially for large-scale HSIs. However, the success of L2GCC shows the broad prospect of unsupervised clustering methods applied to HSI classification. In the future, we will improve HSI pre-processing method and focus on exploring more forms of GCN-based self-supervised autoencoder (e.g., GCN-GAN), and so on.

Table 8.7 Ablation study results on IP, Salinas, UH2013

Name		L2GCC-v_1	L2GCC-v_2	L2GCC-v_3	L2GCC-v_4	L2GCC-v_5	L2GCC-v_6	L2GCC-v_7	L2GCC
Module	PRT	–	✓	–	–	–	✓	✓	✓
	Low-pass GCN	–	–	✓	–	✓	–	✓	✓
	LGAT	–	–	–	✓	✓	✓	–	✓
Dataset	IP	59.76	65.23	64.26	63.28	66.82	69.35	72.39	**73.51**
	Salinas	66.38	72.34	71.72	70.61	73.39	78.23	82.26	**83.15**
	UH2013	54.19	59.73	59.81	53.33	55.17	58.76	62.75	**64.12**

The OA indexes (%) is reported. The pixel to region transformation module (HSI pre-processing) is simplified as “PRT”

References

1. H. Liu, W. Li, X. Xia, M. Zhang, C. Gao, R. Tao, Central attention network for hyperspectral imagery classification. IEEE Trans. Neural Netw. Learn. Syst. **34**, 8989–9003 (2022)
2. Y. Gao, W. Li, M. Zhang, J. Wang, W. Sun, R. Tao, Q. Du, Hyperspectral and multispectral classification for coastal wetland using depthwise feature interaction network. IEEE Trans. Geosci. Remote Sens.Geosci. Remote Sens. **60**, 1–15 (2022)
3. Y. Ding, X. Zhao, Z. Zhang, W. Cai, N. Yang, Graph sample and aggregate-attention network for hyperspectral image classification. IEEE Geosci. Remote Sens. Lett.Geosci. Remote Sens. Lett. **19**, 1–5 (2022). https://doi.org/10.1109/LGRS.2021.3062944
4. Y. Ding, X. Zhao, Z. Zhang, W. Cai, N. Yang, Multiscale graph sample and aggregate network with context-aware learning for hyperspectral image classification. IEEE J. Select. Top. Appl. Earth Observ. Remote Sens. **14**, 4561–4572 (2021). https://doi.org/10.1109/JSTARS.2021.3074469
5. Y. Ding, X. Zhao, Z. Zhang, W. Cai, N. Yang, Y. Zhan, Semi-supervised locality preserving dense graph neural network with ARMA filters and context-aware learning for hyperspectral image classification. IEEE Trans. Geosci. Remote Sens.Geosci. Remote Sens. **60**, 1–12 (2022). https://doi.org/10.1109/TGRS.2021.3100578
6. W. Li, Y. Gao, M. Zhang, R. Tao, Q. Du, Asymmetric feature fusion network for hyperspectral and SAR image classification. IEEE Trans. Neural Netw. Learn. Syst. **34**, 8057–8070 (2022)
7. Y. Zhang, W. Li, R. Tao, J. Peng, Q. Du, Z. Cai, Cross-scene hyperspectral image classification with discriminative cooperative alignment. IEEE Trans. Geosci. Remote Sens.Geosci. Remote Sens. **59**(11), 9646–9660 (2021)
8. Y. Ding, Z. Zhang, X. Zhao, W. Cai, F. He, Y. Cai, Deep hybrid: multi-graph neural network collaboration for hyperspectral image classification. Def. Technol. **23**, 164–176 (2022). https://doi.org/10.1016/j.dt.2022.02.007
9. Y. Cai, Z. Zhang, Z. Cai, X. Liu, X. Jiang, Q. Yan, Graph convolutional subspace clustering: a robust subspace clustering framework for hyperspectral image. IEEE Trans. Geosci. Remote Sens.Geosci. Remote Sens. **59**(5), 4191–4202 (2021)
10. R. Vidal, Subspace clustering. IEEE Sig. Process. Mag. **28**(2), 52–68 (2011)
11. J. Lei, X. Li, B. Peng, L. Fang, N. Ling, Q. Huang, Deep spatialspectral subspace clustering for hyperspectral image. IEEE Trans. Circ. Syst. Video Technol. **31**(7), 2686–2697 (2021)
12. X. Peng, J. Feng, J.T. Zhou, Y. Lei, S. Yan, Deep subspace clustering. IEEE Trans. Neural Netw. Learn. Syst. **31**(12), 5509–5521 (2020)
13. X. Yang, G. Lin, Y. Liu, F. Nie, L. Lin, Fast spectral embedded clustering based on structured graph learning for large-scale hyperspectral image. IEEE Geosci. Remote Sens. Lett. **19**, 1–5 (2022)
14. R. Wang, F. Nie, W. Yu, Fast spectral clustering with anchor graph for large hyperspectral images. IEEE Geosci. Remote Sens. Lett. **14**(11), 2003–2007 (2017)
15. T. Kanungo, D.M. Mount, N.S. Netanyahu, C.D. Piatko, R. Silverman, A.Y. Wu, An efficient k-means clustering algorithm: analysis and implementation. IEEE Trans. Pattern Anal. Mach. Intell. **24**(7), 881–892 (2002)
16. A.Y. Ng, M.I. Jordan, Y. Weiss, On spectral clustering: analysis and an algorithm, in *Proceedings of the Advance Neural Information Processing System (NIPS)* (2002), pp. 849–856
17. T. Yang, C. Lee, S. Yen, Fuzzy objective functions for robust pattern recognition, in *Proceedings of the 2009 IEEE International Conference on Fuzzy Systems* (2009), pp. 2057–2062
18. C.A. Shah, M.K. Arora, P.K. Varshney, Unsupervised classification of hyperspectral data: an ICA mixture model-based approach. Int. J. Remote Sens. **25**(2), 481–487 (2004)
19. Y. Zhang, W. Li, M. Zhang, Y. Qu, R. Tao, H. Qi, Topological structure and semantic information transfer network for cross-scene hyperspectral image classification. IEEE Trans. Neural Netw. Learn. Syst. (2022)
20. J. Lin, C. He, Z.J. Wang, S. Li, Structure preserving transfer learning for unsupervised hyperspectral image classification. IEEE Geosci. Remote Sens. Lett. **14**(10), 1656–1660 (2017)

21. J.M. Murphy, M. Maggioni, Unsupervised clustering and active learning of hyperspectral images with nonlinear diffusion. IEEE Trans. Geosci. Remote Sens. **57**(3), 1829–1845 (2019)
22. E. Elhamifar, R. Vidal, Sparse subspace clustering: algorithm, theory, and applications. IEEE Trans. Pattern Anal. Mach. Intell. **35**(11), 2765–2781 (2013)
23. X. Peng, H. Tang, L. Zhang, Z. Yi, S. Xiao, A unified framework for representation-based subspace clustering of out-of-sample and largescale data. IEEE Trans. Neural Netw. Learn. Syst. **27**(12), 2499–2512 (2016)
24. J. Fan, T. Chen, S. Lu, Superpixel guided deep-sparse-representation learning for hyperspectral image classification. IEEE Trans. Circ. Syst. Video Technol. **28**(11), 3163–3173 (2018)
25. S. Matsushima, M. Brbic, Selective sampling-based scalable sparse subspace clustering, in *Advances in Neural Information Processing Systems*, by eds. H. Wallach, H. Larochelle, A. Beygelzimer, F. d'Alche-Buc, E. Fox, R. Garnett (Curran Associates, Inc., 2019)
26. H. Zhai, H. Zhang, L. Zhang, P. Li, Total variation regularized collaborative representation clustering with a locally adaptive dictionary for hyperspectral imagery. IEEE Trans. Geosci. Remote Sens. **57**(1), 166–180 (2018)
27. S. Chen, D. Zhang, Robust image segmentation using FCM with spatial constraints based on new kernel-induced distance measure. IEEE Trans. Syst. Man Cybern. B **34**(4), 1907–1916 (2004)
28. Y. Ding, Z. Zhang, X. Zhao, D. Hong, W. Li, W. Cai, Y. Zhan, AF2GNN: graph convolution with adaptive filters and aggregator fusion for hyperspectral image classification, in *Information Sciences* (2022)
29. F. Nie, X. Wang, M. Jordan, H. Huang, The constrained Laplacian rank algorithm for graph-based clustering, in *Proceedings of the AAAI Conference on Artificial Intelligence* (2016)
30. J. Xie, R. Girshick, A. Farhadi, Unsupervised deep embedding for clustering analysis, in *Proceedings of the 33rd International Conference on Machine Learning, Series Proceedings of Machine Learning Research*, by eds. M.F. Balcan, K.Q. Weinberger (PMLR, New York, 2016), pp. 478–487
31. Y. Cai, M. Zeng, Z. Cai, X. Liu, Z. Zhang, Graph regularized residual subspace clustering network for hyperspectral image clustering. Inform. Sci. **578**, 85–101 (2021)
32. Y. Cai, Z. Zhang, P. Ghamisi, Y. Ding, X. Liu, Z. Cai, R. Gloaguen, Superpixel contracted neighborhood contrastive subspace clustering network for hyperspectral images. IEEE Trans. Geosci. Remote Sens. **60**, 1–13 (2022)
33. J. Lei, X. Li, B. Peng, L. Fang, N. Ling, Q. Huang, Deep spatial-spectral subspace clustering for hyperspectral image. IEEE Trans. Circ. Syst. Video Technol. **31**(7), 2686–2697 (2021)
34. S. Hou, H. Shi, X. Cao, X. Zhang, L. Jiao, Hyperspectral imagery classification based on contrastive learning. IEEE Trans. Geosci. Remote Sens. (2021)
35. Y. Li, P. Hu, Z. Liu, D. Peng, J.T. Zhou, X. Peng, Contrastive clustering, in *Proceedings of the 2021 AAAI Conference on Artificial Intelligence (AAAI)* (2021), pp. 8547–8555
36. J. Bruna, W. Zaremba, A. Szlam, Y. LeCun, *Spectral Networks and Locally Connected Networks on Graphs* (2013)
37. T. Kipf, M. Welling, *Variational Graph Auto-Encoders* (2016)
38. T.N. Kipf, M. Welling, Semi-supervised classification with graph convolutional networks, in *Proceedings of the ICLR* (2017)
39. P. Veličković, G. Cucurull, A. Casanova, et al., *Graph Attention Networks* (2017)
40. C. Wang, S. Pan, G. Long, X. Zhu, J. Jing, Mgae: marginalized graph autoencoder for graph clustering, in *Proceedings of the 2017 ACM* (2017)
41. J. Park, M. Lee, H.J. Chang, K. Lee, J. Choi, Symmetric graph convolutional autoencoder for unsupervised graph representation learning, in *Proceedings of the IEEE International Conference on Computer Vision* (2019), pp. 6519–6528
42. X. Zhang, H. Liu, Q. Li, X. Wu, Attributed graph clustering via adaptive graph convolution, in *Proceedings of the 28th International Joint Conference on Artificial Intelligence* (AAAI Press, 2019), pp. 4327–4333
43. S. Pan, R. Hu, G. Long, J. Jiang, L. Yao, C. Zhang, Adversarially regularized graph autoencoder for graph embedding. IJCAI **12**, 2609–2615 (2018)

44. R. Horn, C. Johnson, *Matrix Analysis* (Cambridge university press, 2012)
45. J. Xie, R. Girshick, A. Farhadi, Unsupervised deep embedding for clustering analysis, in *ICML* (PMLR, New York, 2016), pp. 478–487
46. B. Scholkopf, A. Smola, K.-R. Muller, Nonlinear component analysis as a kernel eigenvalue problem. Neural Comput. **10**(5), 1299–1319 (1998)
47. R. Achanta, A. Shaji, K. Smith, A. Lucchi, P. Fua, S. Strunk, Slic superpixels compared to state-of-the-art superpixel methods. IEEE Trans. Pattern Anal. Mach. Intell. **34**(11), 2274–2282 (2012)
48. G.E. Hinton, R.R. Salakhutdinov, Reducing the dimensionality of data with neural networks. Science **313**(5786), 504–507 (2006)

SPRINGER NATURE

GPSR Compliance

The European Union's (EU) General Product Safety Regulation (GPSR) is a set of rules that requires consumer products to be safe and our obligations to ensure this.

If you have any concerns about our products, you can contact us on ProductSafety@springernature.com

In case Publisher is established outside the EU, the EU authorized representative is:

Springer Nature Customer Service Center GmbH
Europaplatz 3
69115 Heidelberg, Germany

The manufacturer's authorised representative in the EU is Springer Nature Customer Service Centre GmbH, Europaplatz 3, 69115 Heidelberg, Germany. If you have any concerns regarding our products, please contact ProductSafety@springernature.com

Printed and bound by CPI Group (UK) Ltd, Croydon, CR0 4YY
01/10/2026
02225986-0001